Saturn Building Shell Field Guide

Produced by John Krigger and Chris Dorsi

Illustrated by John Krigger, Bob Starkey,
Steve Hogan, and Mike Kindsfater
This edition compiled by Darrel Tenter
2010 Project Management: Marcus Bianchi (NREL),
Larry Zarker (BPI), and Ed Pollack (DOE)

The *Saturn Building Shell Field Guide outlines* procedures for insulating, air sealing, and improving the shading of existing homes.

The companion volume *Saturn Energy Auditor Field Guide* describes the procedures used to analyze the performance of existing homes.

The companion volume *Saturn HVAC Field Guide* describes the procedures used to evaluate and service forced air heating and cooling systems.

The companion volume *Hydronic Systems Field Guide* includes procedures for evaluating and servicing steam and hot-water space-heating systems.

In compiling this publication, the authors have benefited from the experience of many individuals who have reviewed our documents, related their experiences, or published information from which we've gained insight. Though we can't name everyone to whom we're indebted, we acknowledge the specific contributions of the following people: Martha Benewicz, Michael Blasnik, Anthony Cox, Bruce Manclark, David Miller, Rich Moore, Gary Nelson, Charlie Richardson, Russ Rudy, Russ Shaber, Cal Steiner, Ken Tohinaka, John Tooley, Bill Van Der Meer, and Doug Walter. We take full responsibility, however, for the content and use of this publication.

SATURN
RESOURCE MANAGEMENT

Foreword

The Saturn *Building Shell Field Guide* outlines a set of best practices for technicians who perform air sealing work, install insulation, and replace doors and windows. It also includes guidance on the building shell repairs that are commonly needed as part of these tasks.

The purpose of this guide is to provide specific guidance on building shell retrofits including insulation, air sealing, windows, shading, and reflectivity. This guide incorporates information from standards and specifications published by the Building Performance Institute (BPI) and the Standardized Work Specifications of the Department of Energy's Weatherization Assistance Program (WAP). This guide is also aligned with the International Residential Code (IRC) 2009 and its sister document, the International Energy Conservation Code 2009 (IECC).

The Saturn *Field Guides* have benefitted greatly over the years from generous feedback from our readers. The National Renewable Energy Laboratory (NREL) completed a review of this guide in late 2010 that greatly improved its alignment with DOE and BPI standards along with many helpful technical enhancements. Please help continue this ongoing review process by sending us your comments and suggestions.

John Krigger   Chris Dorsi

jkrigger@srmi.biz   cdorsi@srmi.biz

Chapter 1 outlines the fundamentals covered in this guide which help the shell technician to make a building more energy efficient.

Chapter 2 describes the diagnostic procedures used to evaluate air leakage through the building shell and ducts. These procedures help technicians focus their work where it will be the most beneficial. We've included simple duct tests here because they are often performed at the same time as shell analysis.

Chapter 3 provides guidance on air sealing methods and materials. Since uncontrolled airflow can ruin the performance and durability of any insulation, these procedures should always be performed before insulation is installed.

Chapter 4 covers the installation of insulation. Insulation upgrades are still the best way to improve the performance of existing homes.

Chapter 5 describes the best methods for installing new windows and doors. Though this is a popular retrofit, many installations fall short by allowing air leakage and water intrusion into the home. The details described here should help avoid these common pitfalls.

Chapter 6 includes procedures that are specific to retrofitting mobile homes. These can pay off handsomely especially for the owners of older mobiles.

Chapter 7 focuses on the well-being of both technicians and customers. Health and safety remains paramount to our work in the building trades, and we hope you take to heart the advice contained here.

We use three types of bulleted lists used in this guide.

- A simple bullet denotes a simple list of items.
1. Numbered lists denote a sequence or a specific number of items.
✓ A check mark denotes a checklist of items we recommend accomplishing.

TABLE OF CONTENTS

1: Building Shell Basics

2: Diagnosing Shell & Duct Air Leakage

3: Air Sealing Homes

4: Installing Insulation

5: Windows and Doors

6: Mobile Homes

7: Health and Safety

CHAPTER 1:   BUILDING SHELL BASICS

The purpose of a building shell is to protect occupants from the excesses of temperature and moisture outdoors. Providing acceptable comfort is a goal of modern buildings.

Building shell technicians should understand the basics of comfort and the building shell features that provide comfort at an affordable cost. Building shell technicians must protect their own health and safety and the health and safety of their customer. *See "Health and Safety" on page 207.*

1.1 BASICS OF COMFORT

Residents' perception of indoor comfort is primarily based upon four things: the air temperature, air movement, the radiant temperature, and relative humidity. If one of these factors drifts beyond a comfortable range, residents compensate by adjusting the thermostat to a setting that increases energy consumption.

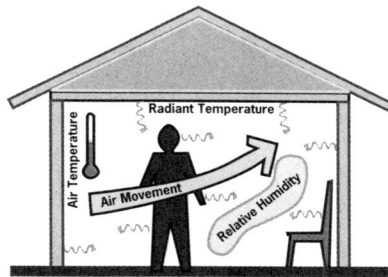

Comfort factors: The perception of comfort is made up of four factors.

1.1.1 Air Temperature

Air temperature, most obvious comfort factor, is the only one that is controlled by your thermostat. In winter, the temperature difference between indoors and outdoors makes heat move across the shell of your home from indoors to outdoors. In summer, the indoor-outdoor difference increases the cost of cooling.

1.1.2 Radiant Temperature.

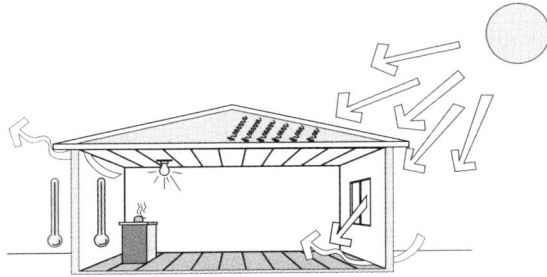

Summer Heat Gains: Solar energy heats homes by shining on roofs, entering through the windows, and heating outdoor air that travels into homes through air leaks. Internal gains, such as appliances and lighting, also release heat into your home.

Your home's radiant temperature is almost as important as its air temperature. Radiant temperature is the temperature of all the objects in the room: the ceilings, floors, and walls and everything else in the rooms. Residents gain or lose heat directly across space to these objects. Cold surfaces hinder comfort in winter and hot surfaces hinder comfort in summer. The radiant temperature of the surfaces determines how much body heat is lost or gained.

R-value is a measure of thermal resistance used in the building and construction industry. In winter, the indoor-outdoor temperature difference and the R-value combine to determine indoor radiant temperature. Windows have the lowest R-value and so they have the lowest winter radiant temperatures.

In summer, the sun tends to heat up your attic, often to as high as 150°F in sunny regions. Good shading keeps radiant temperatures low, promoting good comfort and low air conditioning costs. The temperature of the drywall over head is directly proportional to shading and the R-value of attic insulation.

1.1.3 Air Movement

In winter you perceive moving air in your home as a cold draft. In drafty homes, residents increase the thermostat setpoint in response. In a drafty home, residents may require a thermostat setting of 72°F. In a tight home, residents may be just as comfortable at 68°F. Air movement makes a large difference in the heating energy consumption.

In summer, moving air helps comfort. With moving air from fans, residents may be as comfortable at 82°F as they are at 78°F with no moving air. Moving air in from outdoors, with or without fans, can cool a home whenever outdoor temperatures are sufficiently cool and dry, especially at night.

1.1.4 Relative Humidity

Relative humidity describes the percentage of water vapor in a particular air mass compared to an air mass at the same temperature that is completely saturated with moisture.

Relative humidity (RH) affects summer comfort because it changes the rate at which moisture evaporates from your skin. Sweat evaporates more quickly in dry weather, and that's why 90°F is a more comfortable temperature in Tucson than it is in New Orleans.

Air conditioning helps control indoor relative humidity during the summer by removing moisture from indoor air. However, air conditioning is an relatively expensive way to control indoor relative humidity. Controlling sources of moisture, such as ground water, plumbing leaks, or plant watering, is important for minimizing air-conditioning energy use. *See "Preventing Moisture Problems" on page 222.*

Controlling relative humidity is important to avoid condensation on windows and other cold surfaces during cold weather but relative humidity isn't a major factor in winter indoor comfort.

Controlling relative humidity is very important for minimizing dust mites and mold, which are major sources of allergy and asthma.

1.2 Thermal Flaws in Buildings

Existing buildings have a variety of thermal flaws caused by inadequate design, faulty construction and neglected maintenance. These flaws lead to excessive energy consumption for heating and cooling because they allow excessive heat to travel through the building shell. The thermal weaknesses fall into three (3) broad categories.

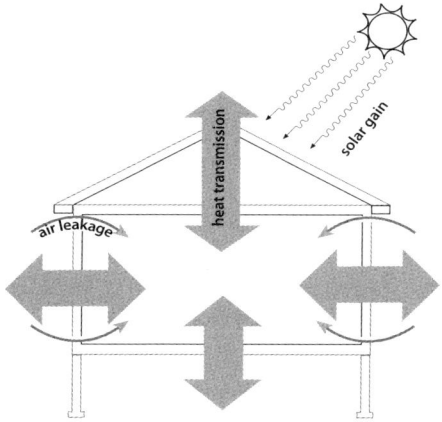

Three major factors in providing comfort: The ideal home has optimal thermal resistance, an airtight air barrier, and orientation and solar control to use solar heat when needed and stop it when not needed.

✓ Inadequate thermal resistance of the building shell.

✓ Excessive air leakage through the building shell

✓ Lack of shade and inadequate exterior surface reflectance, resulting in excessive summer solar gain through the building shell.

1.3 Shell, Enclosure, and Envelope

Three common terms describe the outermost assemblies of a building. These terms are: shell, enclosure, and envelope. We use shell in this book because it is the simplest and least ambiguous term. Enclosure is also a good term that means the same as shell and is used by many building scientists. Envelope is a common term meaning shell, but envelope is also used to mean a hollow building assembly. These two different meanings for envelope make envelope unsuitable for our purpose of describing the shell of the building.

1.4 DEFINITION: THERMAL BOUNDARY

The thermal boundary is a imaginary line, drawn at or inside the building shell, that is useful for discussing heat flow through the building shell. Ideally, an air barrier together with the thermal resistance or insulation is located at the building's thermal boundary. Insulation is responsible for providing adequate thermal resistance between indoors and outdoors. The air barrier is responsible for providing adequate airflow resistance between indoors and outdoors.

The thermal boundary's location was determined by either the building's designer or by an energy specialist who insulates and air seals the building after it is built.

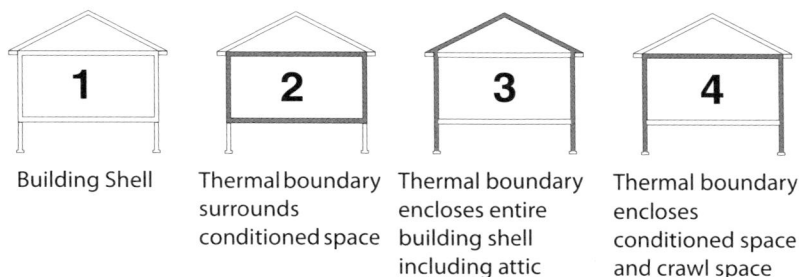

1	2	3	4
Building Shell	Thermal boundary surrounds conditioned space	Thermal boundary encloses entire building shell including attic	Thermal boundary encloses conditioned space and crawl space

Typical building shell and thermal boundary configurations: The building shell is the outermost building assemblies that enclose the building. The thermal boundary is the line where the insulation and air barrier are installed. Examples 3 & 4 typically have incomplete thermal boundaries because of no insulation or air barrier on the ground underneath the building.

The thermal resistance and air barrier work together as a system to limit heat flow across the thermal boundary. The airtightness of the adjacent air barrier has a substantial effect on the building's energy performance. Insulation can retard airflow across the thermal boundary even though insulation's main purpose is to limit heat transmission.

Duct leakage and duct location (either inside the thermal boundary or outside it) are also major energy considerations. If ducts are located outside the thermal boundary, their energy loss is greater than if they are installed inside the thermal boundary.

The thermal boundary is often ambiguous because there is a choice of building assemblies between the conditioned space and outdoors. For example, attics and crawl spaces may be enclosed by the thermal boundary or outside the thermal boundary.

1.5 INSULATION FUNDAMENTALS

Insulation provides buildings with better thermal resistance than the structural materials composing the building's structure. Thermal resistance is most commonly measured by the R-value, which is both a material property and a property of multilayer building assemblies.

Insulation materials have R-values that vary from 1.5 to 6.0 per inch of thickness. Non-insulating materials have lesser R-values per inch. R-values of materials in a building assembly are added to arrive at the total R-value of the assembly. *See "R-values for Common Materials" on page 251.*

Insulation resists heat transmission by conduction, convection, and radiation. Insulation minimizes conduction by trapping air with a minimal amount of fibrous material or by trapping either air or a more thermally resistive gas inside foam bubbles. Insulation resists convection and radiation by filling cavities in building assemblies. *See "Installing Insulation" on page 95.*

1.6 AIR LEAKAGE AND AIR SEALING

Traditional buildings typically have a lot of air leaks. Even most modern buildings don't have any building component specifically designed to be an effective air barrier. An air barrier is a building material or assembly that provides adequate resistance to air leakage between conditioned spaces and outdoors.

Controlling shell air leakage is a key ingredient in a successful energy conservation for buildings. The decisions you make about sealing air leaks affects a building throughout its lifespan.

Air leakage creates the following unpredictable effects and problems in buildings.

- ✓ Air leakage through insulated assemblies reduces the assemblies' thermal resistance.

- ✓ Air leaks create uncomfortable drafts especially in cold weather.

- ✓ Air leakage accounts for a significant percentage of a building's heat loss because the incoming air must be heated or cooled much of the year.

- ✓ Air leakage moves moisture and pollutants into and out of the house.

- ✓ Air leakage causes house pressures that can interfere with the venting of combustion appliances.

1.6.1 Driving Forces for Air Leakage

For air to flow from one zone to another, there must be a pressure difference between the two zones. A zone, as used in this book, is a region separated from other zones by some kind of air barrier. The greater the pressure difference between zones and the leakier the air barrier between zones, the more the air flows between the zones.

Building height and location, weather, and mechanical equipment all cause pressures that drive air leakage through a building. Strong winds may create a positive pressure on one side of a building, and a negative pressure on the opposite side. A chimney or an exhaust fan creates a negative pressure in the building relative to outdoors. A forced air distribution system depressurizes areas near return ducts and pressurizes areas near supply ducts.

Pressure is relative. When we depressurize a building during a blower door test, we create an artificial pressure relative to the outdoors. The outdoors is considered to be zero pressure during this test. If the blower door blows air out of the house, the house has a negative pressure relative to outdoors. We also

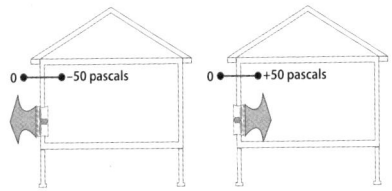

Blower door pressures: A blower door creates pressure differences that allow air leakage to be measured by instruments.

measure pressure between rooms to diagnose air leakage and the driving forces that create air leakage.

Stack Effect and Neutral Pressure Plane

Air can move through a building like gas moving through a chimney. Air tends to enters low in the building (infiltration) and exits at the top of the building (exfiltration) during the heating season. The amount of air entering is equal to the amount of air leaving. This is called the stack effect.

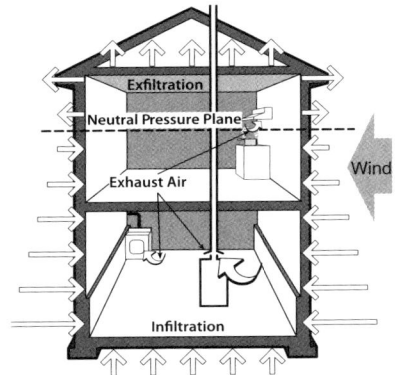

Driving forces for air leakage: Driving sources include stack effect, wind, exhaust fans, and chimneys.

The horizontal plane between the pressurized top and the depressurized bottom of the building is called the neutral pressure plane. Not much air leakage comes in or goes out near the neutral pressure plane. As the building is tightened at the bottom, the neutral pressure plane moves up. As the building is tightened at the top the neutral pressure plane moves down. For the best results, seal air leaks at both the top and bottom of the building. *See "Air Sealing Fundamentals" on page 67.*

Air Sealing and Combustion Safety

Air sealing or duct sealing may affect combustion appliance venting by making house pressures more negative, thereby reducing the available combustion air supply. After all weatherization is complete, technicians must conduct worst-case spillage testing and re-check the safety of all open-combustion appliances. *See "Essential Combustion Safety Tests" on page 207.*

1.6.2 Air Leakage and Ventilation

We depend on air leakage to provide outdoor air for diluting pollutants and admitting fresh air. However, air leakage is unreliable and often brings pollutants into the home, which is why mechanical ventilation works better than air leakage for providing fresh air.

Air leakage serves as ventilation for removing airborne pollutants in traditional buildings. We have learned that preventing pollutants from becoming airborne, eliminating uncontrolled air leakage, and providing controlled mechanical ventilation are the most efficient ways to provide both good comfort and good indoor air quality.

Buildings, equipped with effective air barriers, are designed to be airtight and employ fan-powered ventilation systems. Sealing a building nearly airtight and then providing the correct amount of fan-powered ventilation is the ideal way to maintain good indoor air quality. *See "Other Pollutants" on page 224.*

1.6.3 Air Sealing

The goal of air sealing is to establish a continuous air barrier around all sides of a building. Establishing this effective air barrier in a building requires the following considerations.

- ✓ Sealing big leaks first is the most cost-effective air sealing strategy.

- ✓ Air sealers must choose a strategy to either continue to allow air leakage to ventilate the building or to seal the building to the greatest achievable airtightness and then power ventilate the building.

- ✓ Sealing big leaks is very cost effective and sealing smaller leaks is less cost effective.

- ✓ Sealing a home to a high degree of airtightness creates benefits beyond energy savings such as the ability to use efficient ventilation systems.

- ✓ Heat recovery ventilators and energy recovery ventilators work best in very airtight buildings where all the ventilation air goes through the heat exchanger in the ventilation system, saving heating and cooling energy. *See "Air Sealing Fundamentals" on page 67.*

1.7 SHADING AND REFLECTIVITY FUNDAMENTALS

Solar heating through windows, roof and walls is the leading reason buildings use air conditioning to remove accumulated heat. The ideal building has its long axis parallel with an East-West line. This building orientation maximizes solar heat gain in winter and minimizes solar heat gain in summer (in the Northern Hemisphere).

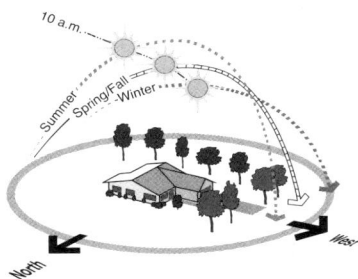

Sun's path relative to home orientation: The ideal home orientation is the long sides of the home facing north and south.

Low-angle sun from the east and west is a heat gain problem for windows facing those directions. A variety of shading options exists to solve these problems starting with special window glass, solar screens, window films, awnings and tree planting. *See "Window Shading" on page 157.* White roofs and walls reflect solar energy and reduce air conditioning energy usage.

CHAPTER 2: DIAGNOSING SHELL & DUCT AIR LEAKAGE

The testing described here helps you analyze the existing air barriers and decide whether and where air sealing or duct sealing is needed.

An air barrier together with the insulation is located at the building's thermal boundary. The airtightness of the adjacent air barrier has a substantial effect on the insulation's effectiveness.

Duct leakage and duct location (either inside the thermal boundary or outside it) are also major energy considerations.

2.1 AIR LEAKAGE AND DUCT LEAKAGE TESTING

Air leakage testing accomplishes a variety of purposes.

- ✓ Measures the home's airtightness level

- ✓ Evaluates the home's ventilation requirements

- ✓ Decides how much time and effort is required to achieve cost-effective air leakage and duct leakage rates, *See page 70.*

- ✓ Determines whether ducts are within or outside the home's thermal boundary

- ✓ Compares the airtightness of the air barriers on either side of the intermediate zone like an attic or crawl space. Comparing the ceiling with the ventilated roof gives the technician an idea of how leaky the ceiling is, for example

- ✓ Evaluates the leakiness of individual air barriers like ceilings

- ✓ Determines the best place to establish the air barrier in an area that has no obvious thermal boundary such as an uninsulated crawl space

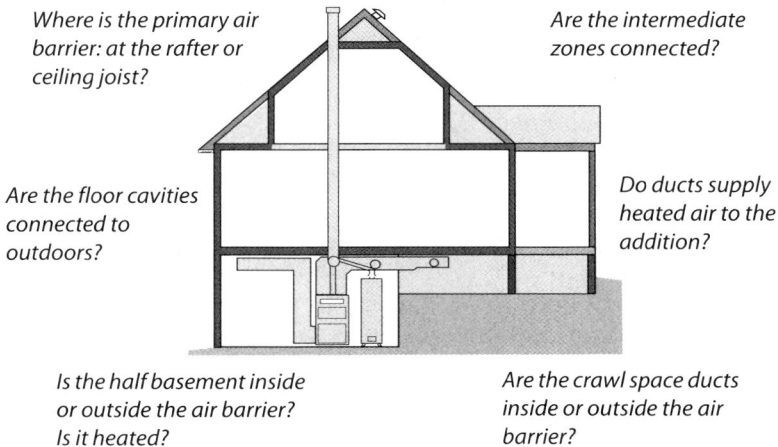

Where is the primary air barrier: at the rafter or ceiling joist?

Are the intermediate zones connected?

Are the floor cavities connected to outdoors?

Do ducts supply heated air to the addition?

Is the half basement inside or outside the air barrier? Is it heated?

Are the crawl space ducts inside or outside the air barrier?

Questions to ask during an energy inspection: Your answers help determine the most efficient and cost-effective location for the air barrier.

The reason for the complexity of air leakage testing is that there is so much uncertainty about air leakage. Testing is needed because there simply is no accurate prescriptive method for determining the severity and location of leaks, especially in complex buildings. Depending on the complexity of a home, you may need to perform varying levels of testing to assess shell and duct leakage.

2.2 Air Sealing and Duct Sealing Strategies

Air sealing and duct sealing are retrofits with a high degree of uncertainty about cost-effectiveness and safety. Unless you measure air leakage and duct leakage, you don't know how much air leakage you're sealing. Both air sealing and duct sealing can make pressures in the combustion zone more negative than before. These retrofits must be accompanied by worst case spillage testing to prevent combustion safety problems. *See "Air Sealing Safety" on page 69.*

2.2.1 Air Sealing Strategies

We believe that there are three basic strategies for reducing air leakage, and at the same time provide adequate ventilation. *See "Other Pollutants" on page 224.*

1. Seal the home as tight as you can get it according to the home performance or weatherization budget. Provide mechanical whole house ventilation using exhaust or supply ventilation fans.

2. Seal the home very airtight during a major home performance retrofit and install a heat or energy recovery ventilator.

3. Seal the home to its minimum ventilation requirement (MVR) and use spot ventilation to remove pollutants at their source. This option is considered obsolete by many building scientists.

Air Sealing with Air Leakage Testing

Dedicate most of your efforts to seal the large air leaks that pass directly through the thermal boundary first. Chasing small leaks isn't worth as much effort.

✓ Perform shell and duct airtightness testing.

✓ Analyze the test results to determine if air sealing is cost-effective.

✓ Locate and seal the air leaks.

✓ Re-test to assess the effectiveness of air sealing efforts.

✓ Stop air sealing when additional air sealing is not cost-effective, or when the minimum ventilation requirement is reached.

2.2.2 Duct Sealing Strategies

The energy impact of duct leakage depends on whether the ducts are located inside or outside of the thermal boundary. If ducts are outside the thermal boundary, they are an extension of the building shell because they contain indoor air. Duct leaks aren't usually a significant energy problem if the ducts are located completely within the thermal boundary. Energy contractors sometimes relocate the thermal boundary to enclose the air handler and ducts for energy savings.

Sealing every joint without testing can be a waste of time and money. Duct leakage tests help you investigate the severity and locations of duct leaks before duct sealing begins. On the other hand, duct testing is time-consuming and doesn't itself save any energy. Therefore prescriptive duct sealing makes sense when the duct leaks are located outside the thermal boundary and easy to find. Use duct airtightness testing as necessary to reduce the uncertainty of your duct sealing work to an acceptable level. Performing the duct leakage test during final inspection is a good compromise between too much testing and no testing at all.

2.3 HOUSE AIRTIGHTNESS TESTING

House airtightness testing was made possible by the development of the blower door. The blower door measures a home's leakage rate at a standard pressure of 50 pascals. This leakage measurement can be used to compare homes with one another and to established air leakage standards.

The blower door also allows the technician to test parts of the home's air barrier to locate air leaks. Sometimes air leaks are obvious. More often, the leaks are hidden, and the technician obtains clues about air leak location from blower door testing. This section outlines the basics of blower door measurement along with some techniques for gathering clues about the location of air leaks.

2.3.1 Blower Door Principles

The blower door creates a 50-pascal pressure difference across the building shell and measures airflow in cubic feet per minute at 50 pascals (CFM_{50}), in order to compare the leakiness of homes. The blower door also creates pressure differences between rooms in the house and intermediate zones like attics and crawl spaces. These pressure differences can give clues about the location and size of a home's hidden air leaks.

Blower Door Components

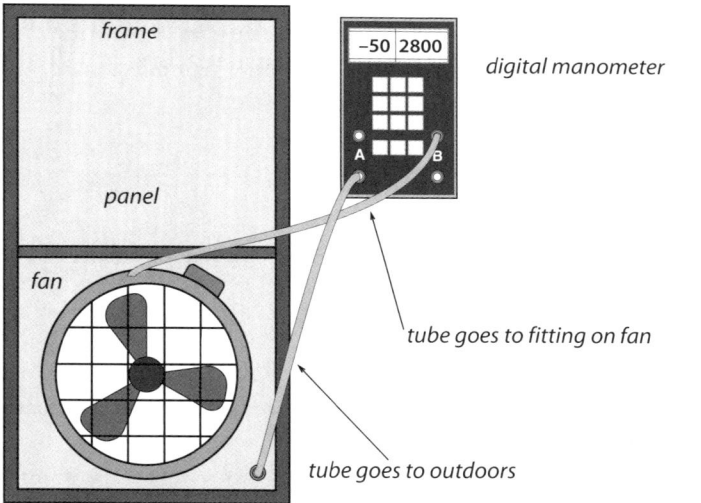

frame

panel

fan

−50 2800

digital manometer

tube goes to fitting on fan

tube goes to outdoors

Digital manometers: Used to diagnose house and duct pressures quickly and accurately.

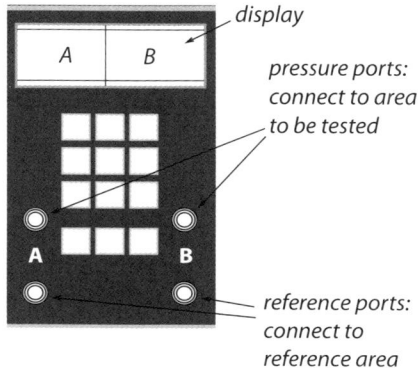

display

A   B

A   B

pressure ports: connect to area to be tested

reference ports: connect to reference area

Blower Door Terminology

Connecting the digital manometer's hoses correctly is essential for accurate testing. There is an accepted method for communicating correct hose connections that helps avoid confusion.

This method uses the phrase *with-reference-to (WRT)*, to distinguish between the input zone and reference zone for a particular measurement. The outdoors is the most commonly used reference zone for blower door testing. The reference zone is considered to be the zero point on the pressure scale.

For example, *house WRT outdoors* = -50 pascals means that the house (input) is 50 pascals negative compared to the outdoors (reference or zero-point). This pressure reading is called the house-to-outdoors pressure difference.

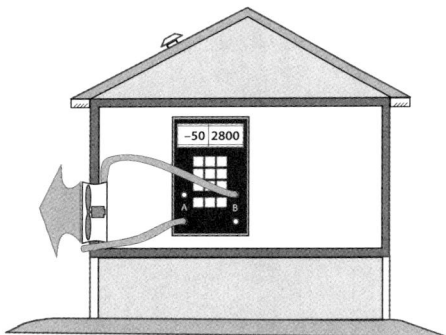

Blower door test: Air barriers are tested during a blower door test, with the house at a pressure of 50 pascals negative with reference to outdoors. This house has 2800 CFM$_{50}$ of air leakage. Further diagnostic tests can help determine where that leakage is coming from.

Low Flow Rings or Range Plates

During the blower door test, the airflow is measured through the fan. This airflow is directly proportional to the surface area of the home's air leaks. For the blower door to measure airflow accurately, the air must be flowing at an adequate speed. Tighter buildings don't have enough air leaks to allow an adequate airspeed to produce the minimum fan pressure of 25 pascals. This low flow condition requires using one of two low flow rings also know as range plates commonly provided with the blower door to reduce the fan's opening and increase air speed, fan pressure, and measurement accuracy.

2.3.2 Preparing for a Blower Door Test

Preparing the house for a blower door test involves putting the house in its normal heating season (or air conditioning) operation with all conditioned zones open to the blower door. Try to anticipate safety problems that the blower door test could cause, particularly with combustion appliances.

- ✓ Identify the location of the thermal boundary and determine which house zones are conditioned.

- ✓ Identify large air leaks that could prevent the blower door from achieving adequate house pressure, such as a pet door flap or evaporative cooler vent.

- ✓ Evaluate situations that may pollute the air or make a mess in the home during a blower door depressurization test — a smoking wood stove or fireplace ashes for example.

- ✓ Put the house into its heating season (or air conditioning) operation with windows, doors, and vents closed and air registers open.

- ✓ Turn off combustion appliances temporarily.

- ✓ Open interior doors so that all indoor areas inside the thermal boundary are connected to the blower door. This could include the basement, conditioned kneewall areas, and closets.

- ✓ Calculate house volume if you plan to use ACH_{50} (air changes per hour at 50 pascals) or ACH_n (natural air changes per hour). Calculate volume by multiplying length by width and then by height to arrive at the number of cubic feet of the home's volume.

Adjusting for Baseline Pressure

To obtain accurate blower door measurements, you must correct your target house pressure reading on the digital manometer to adjust for wind and stack effect. The most common digital manometers have a "Baseline" function that automatically cancels the measurement effect of wind or stack pressures. Follow the manufacturer's instructions about how to adjust the baseline pressure.

Adjust the measured pressure of older or simpler digital manometers by adding or subtracting pressure from the target blower door pressure of 50 pascals. A positive house pressure without blower door operation reduces the target blower door negative house pressure. A negative house pressure would increase the negative target blower door house pressure.

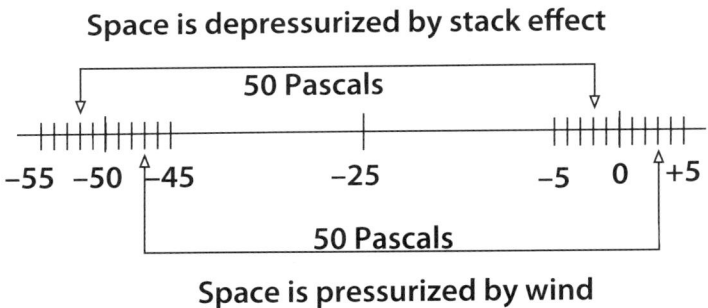

Space is depressurized by stack effect

50 Pascals

−55 −50 −45   −25   −5   0 +5

50 Pascals

Space is pressurized by wind

Block the blower door's opening to prevent ambient airflow through the fan before measuring the baseline pressure. Make sure that the house pressure hose is connected to outdoors and that the fan hose is connected. Measure the house pressure with the blower door off.

If you read a positive house pressure of three pascals with reference to outdoors, add those pascals to -50 pascals and set the house pressure at -47pascals to get your accurate airflow (CFM_{50}). If you read a negative house pressure of 2 pascals subtract those pascals from -50 pascals, and then set the blower door to produce -52pascals to get your accurate airflow.

2.3.3 Blower Door Test Procedures

Follow this general procedure when performing a blower door test.

- ✓ Set up the house for winter conditions with exterior doors, primary windows and storms closed. The door to the basement should be either open or closed, according to whether or not it is considered to be within the thermal boundary.

- ✓ Install blower door frame, panel, and fan in an exterior doorway with a clear path to outdoors. On windy days, install the blower door on the home's leeward side if possible. Pay attention to the blower door's location and any other conditions that may affect test results.

- ✓ Follow manufacturer's instructions for fan orientation and digital manometer setup for either pressurization or depressurization. Depressurization is the most common orientation.

- ✓ Connect Channel A of the digital manometer to measure *house WRT outdoors*. Place the outside hose at least 5 feet away from the fan.

- ✓ Connect Channel B to measure *fan WRT zone near fan inlet*. The zone near the fan inlet is indoors for depressurization and outdoors for pressurization. (Hose must run from the reference port on channel B to outdoors for pressurization.)

- ✓ Adjust for the baseline reading, previously referenced in *"Adjusting for Baseline Pressure" on page 30*. If the manometers used don't zero automatically or manually, the fan pressure must be adjusted to zero this baseline reading (*house WRT outdoors*).

- ✓ Ensure that children, pets, and other potential interferences are at a safe distance from the fan.

- ✓ Turn on the fan and increase its speed slowly until you read 50 pascals of pressure difference between indoors and outdoors.
- ✓ Read the CFM_{50} from channel B of your digital manometer.

Digital Manometer Set-Up Procedures

Follow these instructions for performing a blower door test, using a digital manometer.

- ✓ Turn on the manometer by pushing the ON/OFF button.
- ✓ Push the MODE button twice to set the manometer to read PR/ FL@50.
- ✓ Push the DEVICE button to select a different fan ONLY if not using a Model 3 blower door.
- ✓ With the fan still covered, press the BASELINE button to adjust for the baseline pressure.
- ✓ Push the START button to take baseline readings. Channel B is used as a timer.
- ✓ Push the ENTER button to accept the baseline pressure measurement. ADJ Pa will be displayed on Channel A screen.
- ✓ Remove the No-Flow Plate from the Blower Door Fan. Complete the next two steps for tighter buildings.
  - a. Install the Flow Ring in the Blower Door Fan which matches the expected flow rate. The fan pressure should be at least 25Pa.
  - b. Push CONFIG button until you match the Flow Ring being used.
- ✓ Turn on the blower door fan slowly with the controller. Increase fan speed until the building depressurization on the Channel A screen is between -45 and -55 Pascals. It does not need to be exactly -50 Pascals

✓ The Channel B screen will display the One Point CFM_{50} leakage of the building. If this number is fluctuating a lot, push the TIME AVG button to choose 5 or 10 second averaging or long-term averaging.

Blower Door Test Follow-Up

Be sure to return the house to its original condition.

✓ Inspect combustion appliance pilot lights to ensure that blower door testing did not extinguish them.

✓ Reset thermostats of heaters and water heaters that were turned down for testing.

✓ Remove any temporary plugs that were installed to increase house pressure.

✓ Document any unusual conditions effecting the blower door test and location where the blower door was installed.

2.3.4 Approximate Leakage Area

There are several ways to convert blower door CFM_{50} measurements into square inches of total leakage area. A simple and rough way to convert CFM_{50} into an approximate leakage area (ALA) is to divide CFM_{50} by 10. The ALA can help you visualize the size of openings you're looking for in a home or section of a home.

$$ALA = CFM50 \div 10$$

2.4 EVALUATING HOME VENTILATION LEVELS

Most homes in North America rely on air leakage for ventilation, a practice which is now considered inferior to installing a whole house ventilation system. The American Society of Heating, Refrigeration, and Air Conditioning Engineers (ASHRAE) sets ventilation standards. ASHRAE's current standard is ASHRAE 62.2 2010, which requires whole house ventilation for all but the leakiest homes.

The calculated ventilation rate, evaluated under these ASHRAE standards, is represented by a number of terms. We prefer and use the first term in this list.

- ✓ Minimum ventilation requirement (MVR)
- ✓ Building tightness limit (BTL)
- ✓ Building airflow standard (BAS)
- ✓ Minimum ventilation level (MVL)
- ✓ Minimum ventilation guideline (MVG)

The 2009 International Energy Conservation Code (IECC) requires the ASHRAE 62.2-2010 procedure for sizing whole house ventilation systems. *See "Other Pollutants" on page 224.*

2.4.1 Control of Pollutants

Pollution control should be a very high priority for all homes. The importance of pollution control and whole house ventilation depend on the following.

- ✓ Are sources of moisture like ground water, humidifiers, water leaks, or unvented space heaters causing indoor dampness, high relative humidity, or moisture damage? *See "Moisture Problems" on page 219.*

- ✓ Do occupants complain or show symptoms of building-related illnesses?

✓ Are there combustion appliances in the living space? *See also "Essential Combustion Safety Tests" on page 207.*

✓ Do the occupants smoke?

Pollution Control Tasks

Technicians should survey the home for pollutants before performing air sealing, and perform the following pollutant control measures if needed.

- Repair roof and plumbing leaks.

- Install a ground moisture barrier over any bare soil in crawl spaces or basements.

- Duct dryers and exhaust fans to the outdoors.

- Confirm that combustion appliance vent systems operate properly. Do not air seal homes with unvented space heaters.

- Move paints, cleaning solvents, and other chemicals out of the conditioned space if possible.

The home's occupants have control over the introduction and spread of many home pollutants. Always educate the residents about minimizing pollutants in the home.

2.4.2 ASHRAE 62.2–2010 Ventilation Standard

To comply with ASHRAE 62.2 – 2010, you can use either the formula or the table below to determine the minimum requirements (MVR) in CFM of fan-powered airflow. ASHRAE 62.2-2010 doesn't consider air leakage, as is measured by a blower door in 62-1989, to compute the MVR. Under 62.2-2010, ventilation is provided by a whole house ventilation fan or fans. The MVR is measured in CFM, and this determines the airflow of the required fans.

Follow these steps to determine the MVR under 62.2-2010.

1. Determine the number of occupants by both of the following methods and choose the larger of the two: a) actual number of occupants, or b) number of bedrooms plus one.

2. Determine the floor area of the conditioned space of the home in square feet.

3. Insert these numbers in the formula below, or use *Table 2-1*.

MVR (CFM) = (7.5 cfm x # occupants) + (0.01 x floor area)

Table 2-1: Fan Sizes for Homes with Average Air Leakage

Floor Area (ft²)	No. of Bedrooms				
	0-1	2–3	4–5	6–7	>7
< 1500	30	45	60	75	90
1501–3000	45	60	75	90	105
3001-4500	60	75	90	105	120
4501–6000	75	90	105	120	135
6001–7500	90	105	120	135	150
> 7500	105	120	135	150	165
Fan flow in CFM. From ASHRAE Standard 62.2-2010					

Exceptions to ASHRAE 62.2 – 2010

Whole house ventilation systems aren't required for homes in International Energy Conservation Code (IECC) Zones 3B and 3C, or in homes without mechanical cooling in IECC Zones 1 and 2, or in homes that are conditioned for less than 876 hours per year.

ASHRAE Standard 62.2-2010 allows for a reduction of the whole house ventilation requirements, determined by the formula or table shown previously. The ventilation credit is determined by information contained in ASHRAE Standard 62.2-2010 and ASHRAE Standard 136, A Method of Determining Air Change Rates in Detached Dwellings. These infiltration-credit calculations are incorporated into several software packages, which are available commercially.

Local Ventilation According to ASHRAE 62.2 – 2010

Specify local ventilation for kitchens and bathrooms according to ASHRAE Standard 62.2 – 2007.

1. Specify that bathrooms have a minimum of 50 CFM of intermittent ventilation or 10 CFM of continuous ventilation, supplied by either a bathroom exhaust fan or central ventilator.

2. Specify that kitchens have a minimum of 100 CFM of intermittent ventilation or 5 air changes per hour (ACHn), supplied by either a kitchen exhaust fan or central ventilator.

3. You can reduce the required ventilation rate for kitchens and bathrooms by 20 CFM each, if these rooms have opening windows.

2.4.3 Legacy Minimum Ventilation Requirement (MVR)

According to standard ASHRAE 62-1989, a home must have at least 0.35 natural air changes per hour but not less than 15 CFM per person. ASHRAE considers this standard obsolete and recommends that ASHRAE 62.2 2010 be used instead. The following ASHRAE 62-1989 methodology is included because some energy conservation programs still use this old standard.

The following steps are typical of methods to determine the MVR under 62-1989.

1. Determine the number of occupants by both of the following methods. Choose the larger of the two: a) actual number of occupants, or b) number of bedrooms plus one.

2. Find the home's zone from the map shown here.

3. Decide whether the building is well shielded from wind, "normal", or directly exposed to wind.

4. Find the n-value for the building on the chart. This is where the column representing the building's number of stories meets the row representing its location and the shielding. The n-value converts 50-pascal airflow to natural airflow and vice versa.

5. Calculate the MVR using both of the following formulas. Use the larger number of the two: a) 15 CFM per occupant, or b) 0.35 air changes per hour under natural conditions.

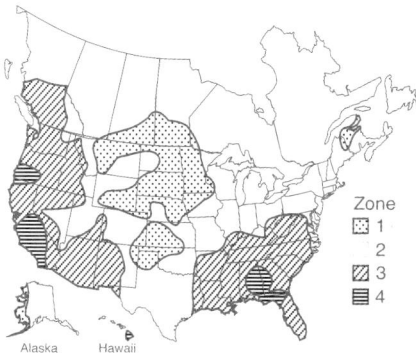

Zone	# of stories ➡	1	1.5	2	3
1	Well-shielded	18.6	16.7	14.9	13.0
	Normal	15.5	14.0	12.4	10.9
	Exposed	14.0	12.6	11.2	9.8
2	Well-shielded	22.2	20.0	17.8	15.5
	Normal	18.5	16.7	14.8	13.0
	Exposed	16.7	15.0	13.3	11.7
3	Well-shielded	25.8	23.2	20.6	18.1
	Normal	21.5	19.4	17.2	15.1
	Exposed	19.4	17.4	15.5	13.5
4	Well-shielded	29.4	26.5	23.5	20.6
	Normal	24.5	22.1	19.6	17.2
	Exposed	22.1	19.8	17.6	15.4

Zone
1
2
3
4

Alaska   Hawaii

Finding the n-value: Find your zone from the map. Pick the correct column by the number of stories in the building. Then decide how exposed the building is and find the n-value.

$$MVR\ (CFM_{50}) = 15\ cfm\ \times\ \#\ occupants\ \times\ n$$

$$MVR\ (CFM_{50}) = \frac{0.35\ \times\ volume\ \times\ n}{60}$$

Mechanical ventilation systems must provide fresh outdoor air when the home's air leakage is below the minimum ventilation requirement (MVR), estimated this way using a blower door.

See "Other Pollutants" on page 224.

2.5 Testing Air Barriers

Leaks in air barriers often cause energy and moisture problems in homes. Air barrier leak testing avoids unnecessary visual inspection and air sealing in hard-to-reach areas.

Blower Door Test: Air barriers are tested during a blower door test, with the house at a pressure of -50 pascals with reference to outdoors. This house has 1800 CFM_{50} of air leakage. Further diagnostic tests can help determine where that leakage is located.

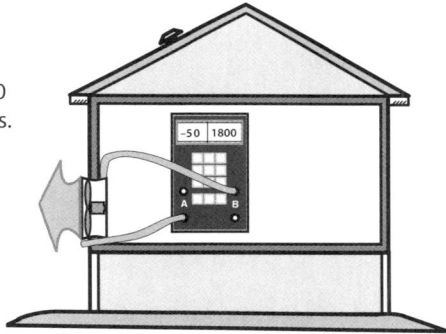

Advanced pressure tests measure pressure differences between zones in order to estimate air leakage between zones. Use these tests to make decisions about where to direct your air sealing efforts.

✓ Evaluate the airtightness of portions of a building's air barrier, especially floors and ceilings.

✓ Decide which of two possible air barriers to air seal — for example, the floor versus foundation walls.

✓ Estimate the air leakage in CFM_{50} through a particular air barrier for the purpose of estimating the effort and cost necessary to seal the leaks.

✓ Determine whether building cavities like floor cavities, porch roofs, and overhangs are conduits for air leakage.

✓ Determine whether building cavities, intermediate zones, and ducts are connected by air leaks.

Air barrier tests provide a range of information from simple clues about which parts of a building are leakiest, to specific estimates of the airflow and hole size through a particular air barrier.

Table 2-2: Building Components and Their Air Permeance

Good air barriers (retrofit) (<1 CFM_{50} per 100 ft.^{2})	Poor air barriers (2–1000 CFM_{50} per 100 ft.^{2})
5/8″ oriented strand board	5/8″ tongue-and-groove wood sheeting
1/2″ drywall	6″ fiberglass batt
4-mil house-wrap paper	1.5″ wet-spray cellulose
asphalt shingles and perforated felt over 1/2″ plywood	wood siding over plank sheathing
1/8″ tempered hardboard	wood shingles over plank sheathing
painted uncracked lath and plaster	blown fibrous insulation

Measurements taken at 50 pascals pressure.
Based on information from: "Air Permeance of Building Materials" by Canada Mortgage Housing Corporation, and estimates of comparable assemblies by the authors.

2.5.1 Choosing an Air Barrier

Where there are two possible air barriers, the most airtight air barrier is usually the best place to air seal in order to meet airtightness standards. The chosen air barrier should be continuous, sealed at seams, relatively impermeable to airflow, and strong enough to avoid damage during windstorms. Hopefully the building's insulation layer is located adjacent to its tightest air barrier. Install any retrofit insulation as close as possible to your chosen air barrier.

The chosen air barrier should be adjacent to the insulation to ensure the insulation's effectiveness. Testing is important to verify that insulation and the home's tightest air barrier are located next to one another.

Intermediate zones are unconditioned spaces that are sheltered within the exterior shell of the house. Intermediate zones include: unheated basements, crawl spaces, attics, enclosed porches, and attached garages. Intermediate zones can either be included within the home's chosen air barrier or outside it.

2.5.2 Simple Pressure Tests

You can find valuable information about the relative leakiness of rooms or sections of the home during a blower door test. Listed below are five simple methods.

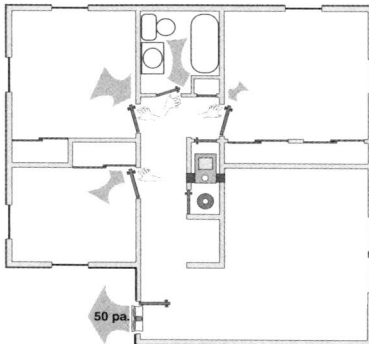

1. *Feeling zone air leakage:* Close an interior door partially so that there is a one-inch gap between the door and door jamb. Feel the airflow along the length of that crack. Compare that airflow intensity with airflow from other rooms using the same technique.

Interior door test: Feeling airflow with your hand at the crack of an interior door gives a tactile sensation of the amount of air leakage coming from the outdoors through that room, which can be compared with other rooms.

2. *Observing the ceiling/attic floor:* Pressurize the home and observe the top floor ceiling from the attic with a good flashlight. Air leaks will show in movement of loose fill insulation, blowing dust, moving cobwebs, etc. You can also use a small piece of tissue paper to disclose air movement.

3. *Observing smoke movement:* Pressurize the home and observe the movement of smoke through the house and out of its air leaks.

4. *Room pressure difference:* Check the pressure difference between a closed room or zone and the main body of a home. Larger pressure differences indicate larger potential air leakage within the closed room or else a tight air barrier between the room and main body. A small pressure difference

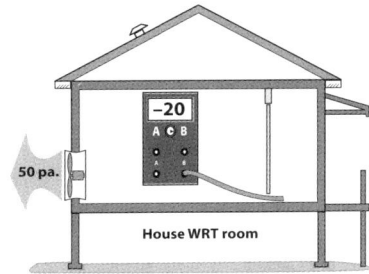

-20

50 pa.

House WRT room

Bedroom test: This pressure difference may be caused by the bedroom's leaky exterior walls or its tight interior walls, separating the bedroom from the main body of the home. This test can also determine whether or not a confined combustion zone is connected to other rooms.

means little leakage to the outdoors through the room or a leaky air barrier between the house and room.

5. *Room airflow difference:* Measure the house CFM_{50} with all interior doors open. Close the door to a single room, and note the difference in the CFM_{50} reading. The difference is the approximate leakage through that room.

Tests 1, 2, and 3 present good customer education opportunities. Feeling airflow or observing smoke are simple observations, but have helped identify many air leaks that could otherwise have remained hidden.

When airflow within the home is restricted by closing a door, as in tests 4 and 5, it may take alternative indoor paths that render these tests somewhat inaccurate. Only practice and experience can guide your decisions about the applicability and usefulness of these general indicators.

2.5.3 Using a Manometer to Test Air Barriers

A manometer, used for blower door testing, can also measure pressures between the house and its intermediate zones during blower door tests.

The blower door, when used to create a house-to-outdoors pressure of -50 pascals, also creates house-to-zone pressures of between 0 and -50 pascals in the home's intermediate zones. The amount of depressurization depends on the relative leakiness of the zone's two air barriers; one air barrier between the house and zone, and one between the zone and outdoors.

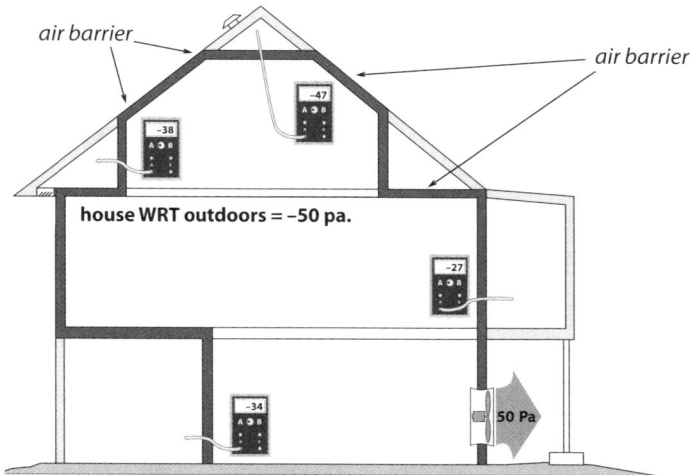

Pressure testing building zones: Measuring the pressure difference across the assumed air barrier tells you whether the air barrier and insulation are aligned. If the manometer reads close to -50 pascals, they are aligned, assuming the tested zones are well connected to outdoors.

For example, in an attic with a fairly airtight ceiling and a well ventilated roof, the attic will indicate that it is mostly outdoors by having a house-to-zone pressure of -45 to -50 pascals. The leakier the ceiling and the tighter the roof, the smaller the negative house-to-zone pressure will be. This holds true for other intermediate zones like crawl spaces, attached garages, and unheated basements.

Zone Leak Testing Methodology

Depressurize house to -50 pascals with a blower door.

1. Find an existing hole or drill a hole through the floor, wall, or ceiling between the conditioned space and the intermediate zone.

2. Connect the reference port (digital manometer) or the low pressure port (analog manometer) to a hose connected into the zone.

3. Leave the input port (digital manometer) or the high pressure port (analog manometer) open to the indoors.

4. Read the negative pressure given by the manometer. This is the house-to-zone pressure which is -50 pascals if the air barrier between house and zone is airtight and the zone is open to outdoors.

5. If the reading is significantly less negative than -45 pascals, find the air barrier's largest leaks and seal them.

6. Repeat steps 1 through 5, performing more air sealing as necessary, until the pressure is as close to -50 pascals as possible.

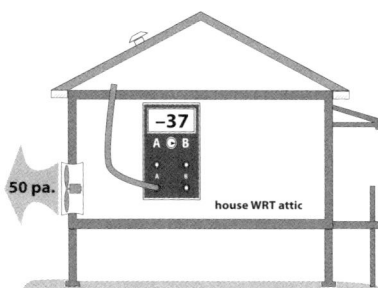

House-to-attic pressure: This commonly used measurement is convenient because it requires only one hose.

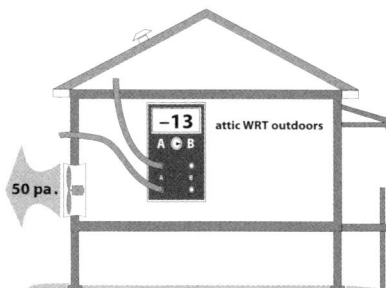

Attic-to-outdoors pressure: This measurement confirms the one to the left because the two add up to -50 pascals.

Leak Testing Building Cavities

Building cavities such as wall cavities, floor cavities between stories, and dropped soffits in kitchens and bathrooms can also be tested as described above to determine their connection to the outdoors as shown here.

Testing Zone Connectedness

Sometimes it is useful to determine whether two zones are connected by an air passage like a large bypass. Measuring the house-to-zone pressure during a blower door test before and then after opening the other zone to the outdoors can establish whether the two zones are connected. You can also open an interior door to one of the zones and check for pressure changes in the other.

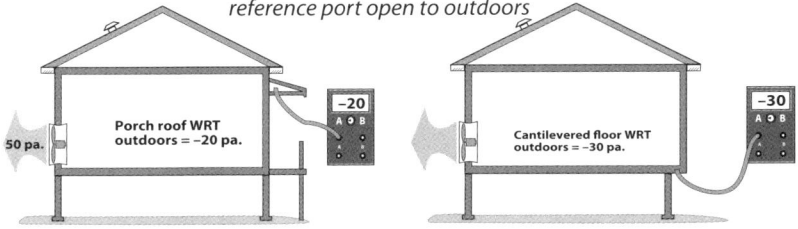

These examples assume that the manometer is outdoors with the reference port open to outdoors

Porch roof WRT outdoors = −20 pa.

−20
A ● B

50 pa.

Cantilevered floor WRT outdoors = −30 pa.

−30
A ● B

Porch roof test: If the porch roof were outdoors, the manometer would read near 0 pascals. We hope that the porch roof is outdoors because it is outside the insulation. We find, however, that it is partially indoors, indicating that it may harbor significant air leaks through the thermal boundary.

Cantilevered floor test: We hope to find the cantilevered floor to be indoors. A reading of -50 pascals would indicate that it is completely indoors. A reading less negative than -50 pascals is measured here, indicating that the floor cavity is partially connected to outdoors.

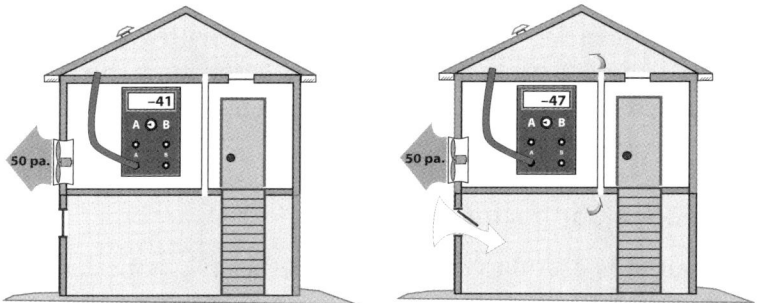

−41
A ● B

50 pa.

−47
A ● B

50 pa.

Zone connectedness: The attic measures closer to outdoors after the basement window is opened, indicating that the attic and basement are connected by a large bypass.

2.5.4 Locating the Primary Air Barrier

Zone pressures are one of several factors used to determine where the thermal boundary should be. Where to air seal and where to insulate are necessary retrofit decisions. When there are two choices of where to insulate and air seal, zone pressures along with other considerations help you decide.

For zone leak testing, you can use the house-to-zone pressure to determine which of two air barriers is the tighter air barrier or primary air barrier.

✓ Readings of negative 25-to-50 pascals house-to-attic pressure mean that the ceiling is tighter than the roof. If the roof is quite airtight, achieving a 50-pascal house-to-attic pressure difference may be difficult. However if the roof is well ventilated, achieving a near -50-pascal difference should be possible.

✓ Readings of negative 0-to-25 pascals house-to-attic pressure mean that the roof is tighter than the ceiling. If the roof is well-ventilated, the ceiling has even more leak area than the roof's vent area.

✓ Readings around -25 pascals house-to-attic pressure indicate that the roof and ceiling are equally airtight or leaky.

Pressure readings more negative than -45 pascals indicate that the primary air barrier is adequately airtight. Less negative pressure readings indicate that air leaks should be located and sealed. Pressure measurements from this test only give a relative indication of air leakage. for a more absolute measurement of air leakage see, *"Add-a-Hole Zone Leakage Measurement" on page 49.*

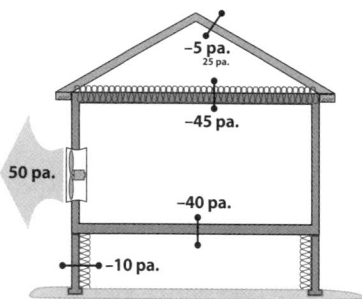

−5 pa.
25 pa.

−45 pa.

50 pa.

−40 pa.

−10 pa.

Pressure measurements and air-barrier location: The air barrier and insulation are aligned at the ceiling. The crawl space pressure measurements show that the floor is the primary air barrier and the insulation is misaligned — installed at the foundation wall. We could decide to close the crawl space vents and air seal the crawl space. Then the insulation would be aligned with the air barrier.

Floor Versus Crawl Space

The floor shown here is tighter than the crawl space foundation walls. If the crawl space foundation walls are insulated, holes and vents in the foundation wall should be sealed until the pressure difference between the crawl space and outside is as negative you can make it – ideally more negative than -45 pascals. A leaky foundation wall renders its insulation nearly worthless.

If the floor above the crawl space were insulated instead of the foundation walls in the above example, the air barrier and the insulation would be aligned.

If a floor is already insulated, it makes sense to establish the air barrier there. If the foundation wall is more airtight than the floor, that would be one reason to insulate the foundation wall.

Attic Boundary

Generally, the thermal boundary (air barrier and insulation) should be between the conditioned space and attic. An exception would be insulating the roof to enclose an attic air handler and its ducts within the thermal boundary.

Garage Boundary

The air barrier should always be between the conditioned space and a tuck-under or attached garage to separate the living spaces from this unconditioned and often polluted zone.

Duct Location

The location of ducts either within or outside the thermal boundary is an important factor in determining the cost-effectiveness of duct sealing and insulation. Many retrofitters now prefer including the heating ducts within the thermal boundary because it reduces energy waste from duct leakage.

2.5.5 Add-a-Hole Zone Leakage Measurement

If you are still unsure of the location and severity of air leaks after the simpler diagnostic tests, you can use this add-a-hole procedure to estimate the actual airflow between the house and attic.

Based on the original house-to-attic pressure you measure, *Table 2-3* allows you to choose one of three pressure drops (5, 10, and 15 pascals) that you create by opening a hole by way of an attic hatch or other opening between the house and attic. You estimate the area of the opening required to achieve that pressure drop. *Table 2-3* then provides a multiplier to convert square inches of opening to CFM_{50}. After determining this CFM_{50} leakage figure, you can determine the percentage of the home's air leakage coming through the attic.

Follow this procedure.

1. Establish hoses into attic (or other zone) for measuring house WRT attic.

2. Depressurize house to -50 pascals.

3. Record the house's CFM_{50}.

4. Measure and record the house-to-attic pressure (or other zone pressure). Locate that house-to-zone pressure in the H/Z column of *Table 2-3 on page 51.* For each initial pressure there are values for CFM_{50} per square inch of opening for three specific pressure drops, 5, 10, and 15 pascals.

5. Reduce the pressure you found in Step 4 by 5, 10, or 15 pascals by opening the attic hatch while increasing the blower door's speed to maintain -50 pascals house-to-outdoors pressure difference. When you reach your target house-to-attic pressure, make sure the house-to-outdoor pressure is still -50 pascals.

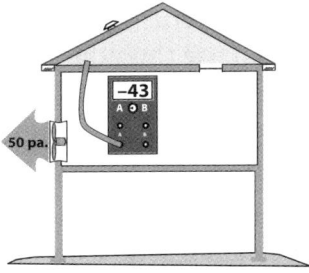

Add-a-hole test: The first house-to-attic pressure is -43 pascals. If we drop the pressure by 10 pascals, every square inch of opening will represent 7 CFM_{50} of leakage between house and attic. *See Table 2-3 on page 51.*

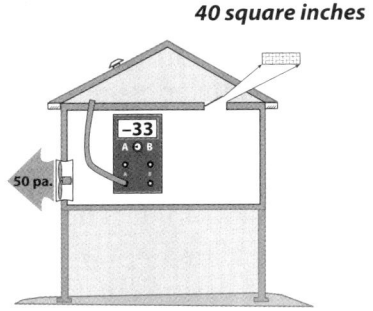

Add-a-hole test 2: Opening a hole of approximately 40 square inches drops the second house-to-zone pressure by 10 pascals.

OPENING AREA (SQ.IN.) X FACTOR (CFM_{50}/SQ.IN.) = CFM_{50}

40 SQ.IN. X 7 CFM_{50}/SQ.IN. = 280 CFM_{50}

6. Estimate the area of the opening you made in Step 5. Multiply this opening's estimated area in square inches times the factor in *Table 2-3 on page 51* to find the CFM_{50} leaking between house and attic.

Table 2-3: Add-a-Hole CFM_{50} per Square Inch of Hole

Measured Pressure in Pascals		Pressure Drop			Measured Pressure in Pascals		Pressure Drop		
H/Z^a	Z/O^b	5 Pa	10 Pa	15 Pa	H/Z^a	Z/O^b	5 Pa	10 Pa	15 Pa
49	1	3	1.7	1.1	35	15	19	9	6.8
48	2	5	2.8	1.9	34	16	19	9	6.9
47	3	7	4	2.5	33	17	19	9	7.0
46	4	9	5	3.1	32	18	19	9	7.1
45	5	11	6	3.6	31	19	20	9	7.1
44	6	12	6	4.1	30	20	20	9	7.1
43	7	13	7	4.5	29	21	20	8	7.0
42	8	14	7	4.9	28	22	19	8	7.0
41	9	15	7	5.3	27	23	19	8	6.8
40	10	16	8	5.6	26	24	18	8	6.7
39	11	17	8	5.9	25	25	18	8	6.5
38	12	17	8	6.2	20	30	15	6	--
37	13	18	8	6.5	15	35	10	3	--
36	14	18	8	6.6	10	40	5	--	--

Courtesy of Michael Blasnik

a. House-to-Zone pressure difference
b. Zone-to-Outdoors pressure difference

2.5.6 Open-a-Door Zone-Leakage Measurement

The open-a-door method is a another way of determining how much leakage in CFM_{50} travels through an intermediate zone like a walk-up attic, basement, or attached garage. This method requires a door between the house and the intermediate zone. Use your programmable calculator to perform the calculations. If you don't have a calculator, use the tables provided here.

1. Perform a blower door test and measure CFM_{50} with the door between the house and intermediate zone closed.

2. During the test, measure the pressure difference between the house and zone.

3. Open the door between house and zone and measure CFM_{50} again. Also measure the pressure difference between the house and zone. It should now be 0 pascals.

4. Find the exterior leakage factor from *Table 2-4 on page 54*, and multiply the CFM_{50} difference between door-open and door-closed blower door tests by this factor.

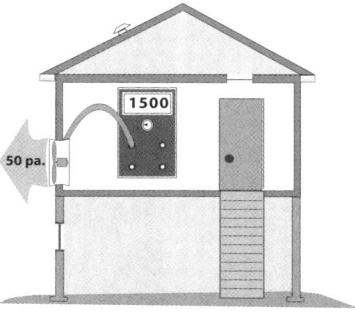

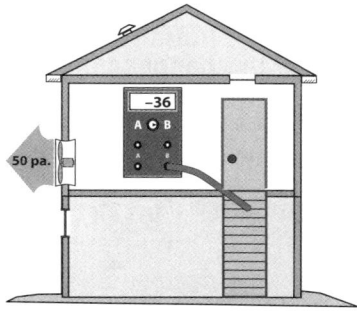

Open-a-door test: Start with a CFM_{50}
reading and a pressure difference between house and the basement zone.

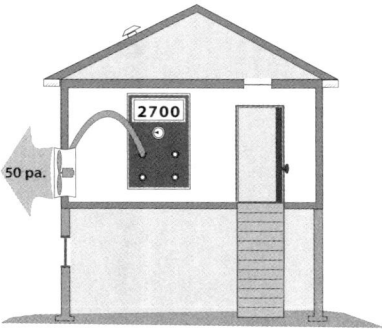

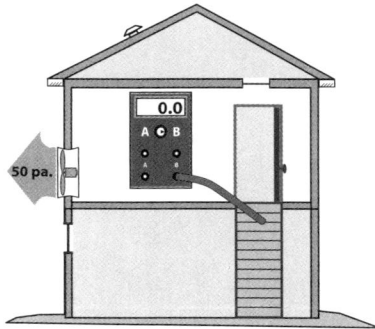

Open-a-door test 2: Now open the door, and read the new CFM_{50}, while making sure that there is no pressure difference across the door.

CFM_{50} **DIFFERENCE X LEAKAGE RATIO =**

CFM_{50} **(HOUSE/ZONE OR ZONE/EXTERIOR)**

1200 CFM_{50} X .96 = 1150 CFM_{50}

Calculation for house-to-zone leakage

1200 CFM_{50} X 1.78 = 2140 CFM_{50}

Calculation for zone-to-outdoors leakage

Table 2-4: Open-a-Door CFM_{50} Leakage

(Per CFM_{50} change upon opening)

House Pressures		Leakage Ratios		House Pressures		Leakage Ratios	
H/Z	Z/O	Int Lk	Ext Lk	H/Z	Z/O	Int Lk	Ext Lk
48	2	.14	1.14	27	23	2.28	2.53
47	3	.19	1.19	26	24	2.5	2.64
46	4	.25	1.24	25	25	2.77	2.76
45	5	.31	1.29	24	26	3.05	2.89
44	6	.37	1.34	23	27	3.41	3.04
43	7	.43	1.39	22	28	3.73	3.19
42	8	.49	1.44	21	29	4.16	3.36
41	9	.56	1.49	20	30	4.61	3.54
40	10	.63	1.54	19	31	5.2	3.76
39	11	.70	1.6	18	32	5.78	3.98
38	12	.78	1.66	17	33	6.58	4.24
37	13	.87	1.72	16	34	7.38	4.52
36	14	.96	1.78	15	35	8.5	4.87
35	15	1.06	1.85	14	36	9.63	5.21
34	16	1.17	1.91	13	37	11.3	5.68
33	17	1.3	1.98	12	38	12.9	6.14
32	18	1.42	2.06	11	39	15.6	6.79
31	19	1.54	2.12	10	40	18.3	7.43
30	20	1.71	2.23	9	41	22.9	8.4
29	21	1.99	2.31	8	42	27.5	9.37
28	22	2.07	2.42				

Courtesy of Michael Blasnik

2.5.7 Decisions about Basement and Crawl Spaces

The importance of creating an effective air barrier at the foundation walls or floor depends on how much of the home's air leakage is coming through the foundation or floor.

The retrofitter may choose either the first floor or the foundation wall as the air barrier. If installing insulation in that building component is also a weatherization priority, the insulation is installed in either the floor or the foundation wall, depending on which was chosen as the air barrier. Most basements and crawl spaces in existing homes are uninsulated.

The results of air barrier tests are only one deciding factor in selecting the thermal boundary's location. Moisture problems, duct and furnace location, and the necessity of crawl-space venting are other important considerations.

Basement insulation may not be a very practical weatherization option because of moisture concerns, cost, or the need to drywall and tape any newly insulated interior surfaces. Crawl space insulation poses fewer problems and is often undertaken using foam sheeting, wet-spray cellulose or fiberglass, spray two-part foam, or even foil-faced (FSK) fiberglass.

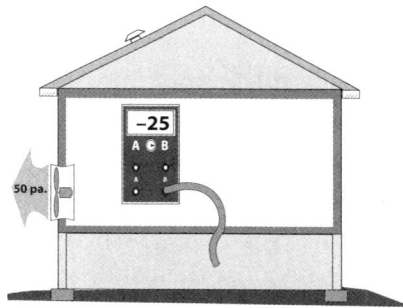

House-to-crawl-space pressure: Many homes with crawl spaces have an ambiguous thermal boundary at the foundation. Is the air barrier at the floor or foundation wall? Answer: in this case, each of the two building assemblies is equally leaky or airtight.

The tables presented here summarize the decision factors for choosing between the floor and the foundation wall as the air barrier. You may also encounter situations that aren't addressed here.

When a home has a basement and crawl space connected together, both *Table 2-5 on page 56* and *Table 2-6 on page 57* are relevant to the decision making process of selecting the air barrier and location for insulation if insulation is found to be cost-effective. A basement may even be divided from its adjoining crawl space to allow the basement to be within the air barrier and the crawl space to be outside the air barrier.

Table 2-5: Crawl Space: Where Is the Air Barrier?

Factors favoring foundation wall	Factors favoring floor
Ground moisture barrier and good perimeter drainage present or planned	Damp crawl space with no improvement offered by weatherization
Foundation walls test tighter than floor	Floor air sealing and insulation are reasonable options, considering access and obstacles
Vents can be closed off	Floor tests tighter than foundation walls
Furnace, ducts, and water pipes located in crawl space	No furnace or ducts present
Concrete or concrete block walls are easily insulated	Building code or code official forbids closing vents
Floor air sealing and insulation would be more difficult than sealing and insulating the foundation	Rubble masonry foundation wall
Foundation wall is already insulated	Floor is already insulated

Table 2-6:   Unoccupied Basement: Where Is the Air Barrier?

Favors foundation wall	Favors floor
Ground drainage and no existing moisture problems	Damp basement with no solution during weatherization
Interior stairway between house and basement	Floor air sealing and insulation is a reasonable option, considering access and obstacles
Ducts and furnace in basement	No furnace or ducts present
Foundation walls test tighter than the floor	Floor tests tighter than foundation walls
Basement may be occupied some day	Exterior entrance and stairway only
Laundry in basement	Rubble masonry foundation walls
Floor air sealing and insulation would be very difficult	Dirt floor or deteriorating concrete floor
Concrete floor	Badly cracked foundation walls

2.6 DUCT AIRTIGHTNESS TESTING

Duct testing is included in this chapter because this procedure is often performed at the same time as shell pressure diagnostics. Additionally, if the ducts are located outside the shell – as in an unfinished attic, crawl space, or garage – the ducts are in fact a part of the building shell.

The most accurate method of measuring duct leakage is with a duct blower. That advanced procedure is described in the *Saturn HVAC Systems Field Guide*. The following pressure pan method is performed in conjunction with a blower door test. Though less accurate than a duct blower test, it can give technicians a quick assessment of duct airtightness.

2.6.1 Pressure Pan Testing

Pressure pan tests can help identify leaky or disconnected ducts. With the house depressurized by the blower door to -25 or -50 pascals with reference to outdoors, pressure pan readings are taken at each supply and return register.

Pressure pan testing is reliable for mobile homes and small site-built homes where the ducts are outside the air barrier. Pressure pan tests don't accurately disclose duct leakage into the conditioned space.

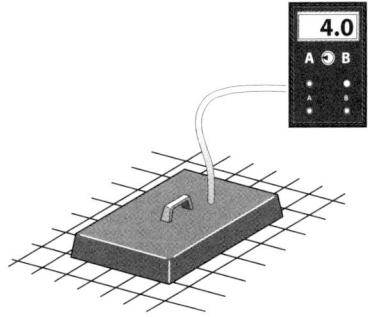

A pressure pan: Blocks a single register and measures the air pressure behind it, during a blower door test. The magnitude of this pressure is an indicator of duct leakage.

Basements are often included in the conditioned living space of a home. In these homes, pressure pan testing isn't necessary, although air sealing the return ducts may be needed to protect against depressurization. But if the basement is accessed from the outside and rarely used, the basement may be considered outside the conditioned living space. In this case, open a window or door between the basement and outdoors, and close any door or hatch between conditioned spaces and basement during pressure pan testing.

Follow this procedure to perform a pressure pan test.

1. Install blower door and set-up house for winter conditions. Open all interior doors.

2. If the basement is conditioned living space, open the door between basement and upstairs living spaces. If the basement is considered outside the conditioned living space, close the door between basement and upstairs living spaces and open a basement window.

3. Turn furnace off. Remove furnace filter, and tape filter slot if one exists. Ensure that all grilles, registers, and dampers are fully open.

4. Temporarily seal any outside fresh air intakes to the duct system. Seal supply registers in zones that are not intended to be heated – an uninhabited basement or crawl space, for example.

5. Open attics, crawl spaces, and garages as much as possible to the outside so they don't create a secondary air barrier.

6. Connect a hose between pressure pan and the input tap on the digital manometer. Leave the reference tap open.

7. With the blower door at -25 pascals, place the pressure pan completely over a grille or register to form a tight seal. Record the reading, which should be a positive pressure.

8. If a grille is too large or a supply register is difficult to access (under a kitchen cabinet, for example), seal the grille or register with masking tape. Insert a pressure probe through the masking tape and record reading.

9. Repeat this test for each register and grille in a systematic fashion.

Pressure Pan Duct Standards

If the ducts have no leakage to the outside, no pressure difference (0 pascals) will be measured during a pressure-pan test. The higher the measured pressure pan reading, the more connected the duct is to the outdoors. Readings greater than 1.0 pascal require investigation and sealing of leaks causing the reading.

Pay particular attention to registers connected to ducts that are located in areas outside the conditioned living space. These spaces include attics, crawl spaces, garages, and unoccupied

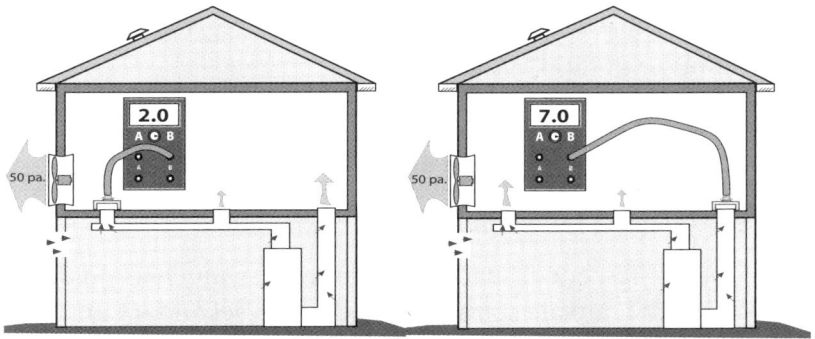

Pressure pan test: A pressure pan reading of 2 indicates moderate duct air leakage.

Problem register: A pressure reading of 7 pascals indicates major air leakage near the tested register.

basements as described previously. Also test return registers attached to stud cavities or panned joists used as return ducts. Leaky ducts located outside the conditioned living space may show pressure pan readings of up to 25-to-50 pascals if they have large leaks.

2.6.2 Measuring Duct Air Leakage with a Duct Blower

Pressurizing the ducts with a duct blower measures total duct leakage. The duct blower is the most accurate common measuring device for duct air leakage. It consists of a fan, a digital manometer or set of analog manometers, and a set of reducer plates for measuring different leakage levels. Using a blower door with a duct blower measures leakage to outdoors.

Measuring Total Duct Leakage

The total duct leakage test measures leakage to both indoors and outdoors. The house and intermediate zones should be open to the outdoors by way of windows, doors, or vents. Opening the intermediate zones to outdoors insures that the duct blower is measuring only the ducts' airtightness — not the airtightness of ducts combined with other air barriers like roofs, foundation walls, or garages.

Supply and return ducts can be tested separately, either before the air handler is installed in a new home or when an air handler is removed during replacement.

Follow these steps when performing a duct airtightness test.

1. Install the duct blower in the air handler or to a large return register, either using its connector duct or simply attaching the duct blower itself to the air handler or return register with cardboard and tape.

2. Remove the air filter(s) from the duct system.

3. Seal all supply and return registers with masking tape or other non-destructive sealant.

4. Open the house, basement or crawl space, containing ducts, to outdoors.

5. Drill a $^1/_4$ or $^5/_{16}$-inch hole into a supply duct a short distance away from the air handler and insert a manometer hose. Connect a manometer to this hose to measure *duct with reference to (WRT) outdoors*. (Indoors, outdoors, and intermediate zones should ideally be opened to each other in this test).

6. Connect an airflow manometer to measure *fan WRT the area near the fan*.

Check manometer(s) for proper settings. Digital manometers require your choosing the correct mode, range, and fan-type settings.

1. Turn on the duct blower and pressurize the ducts to 25 pascals.

2. Record duct-blower airflow.

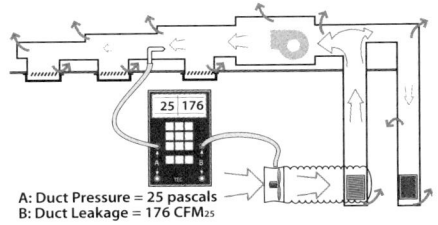

A: Duct Pressure = 25 pascals
B: Duct Leakage = 176 CFM$_{25}$

3. While the ducts are pressurized, start at the air handler and move outward feeling for leaks in the air handler, main ducts, and branches.

Total duct air leakage measured by the duct blower: All registers are sealed except the one connecting the duct blower to the system. Pressurize the ducts to 25 pascals and measure airflow.

4. After testing and associated air-sealing are complete, restore filter(s), remove seals from registers, and check air handler.

Measuring Duct Leakage to Outdoors

Measuring duct leakage to outdoors gives you a duct-air-leakage value that is directly related to energy waste and the potential for energy savings.

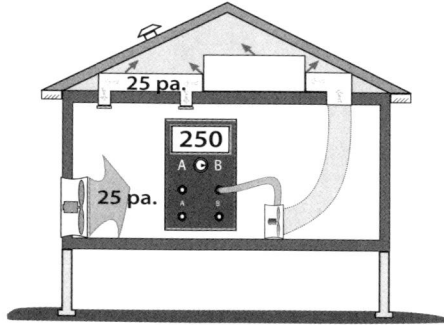

Measuring duct leakage to outdoors: Using a blower door to pressurize the house with a duct blower to pressurize the ducts measures leakage to the outdoors — a smaller number and a better predictor of energy savings. This test is the preferred for evaluating duct leakage for specialists in both shell air leakage and duct air leakage whenever a blower door is available.

1. Set up the home in its typical heating and cooling mode with windows and outside doors closed. Open all indoor conditioned areas to one another.

2. Install a blower door, configured to pressurize the home.

3. Connect the duct blower to the air handler or to a main return duct.

4. Pressurize the ducts to +25 pascals by increasing the duct blower's speed until this value is reached.

5. Pressurize the house until the pressure difference between the house and duct is 0 pascals (*house WRT ducts*).

6. Read the airflow through the duct blower. This value is duct leakage to outdoors.

2.7 DUCT-INDUCED HOUSE PRESSURES

House pressures drive air through leaks in the building shell and can cause open combustion appliances to backdraft. For energy conservation and safety, you should measure duct-induced house pressures and try to limit them.

An improperly balanced air handling system can reduce comfort, building durability, and indoor air quality. Duct-induced room pressures can increase air leakage through the building shell from 1.5 to 3 times, compared to when the air handler is off.

The following test measures pressure differences between the main body of the house and outdoors, between each room and outdoors, and between the combustion zone and outdoors. A pressure difference greater than +4.0 pascals or more negative than -4.0 pascals should be corrected. If the pressure imbalance is the result of occupant behavior such as covering supply or return grilles, discuss these issues with the customer.

2.7.1 Dominant Duct Leakage

This test helps determine whether duct sealing efforts should be directed to the supply or return duct system. This test doesn't measure the amount of duct leakage.

1. Set up house for winter conditions. Close all windows and exterior doors. Turn off all exhaust fans.

2. Open all interior doors, including door to basement.

3. Turn on the air handler.

4. Measure the house-to-outdoors pressure difference.

A positive pressure indicates that the return ducts (which pull air from leaky intermediate zones) are leakier than the supply ducts. A negative pressure indicates that the supply ducts (which push air into intermediate zones through their leaks) are leakier than return ducts. A pressure at or near zero indicates equal supply and return leakage or else little duct leakage.

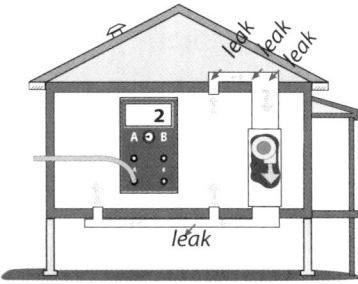

Dominant return leaks: When return leaks are larger than supply leaks, the house shows a positive pressure with reference to the outdoors.

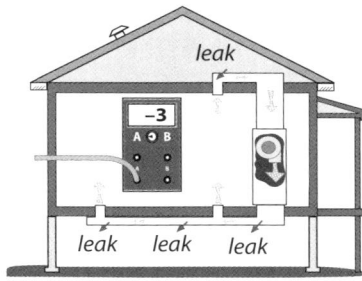

Dominant supply leaks: When supply leaks are larger than return leaks, the house shows a negative pressure with reference to the outdoors.

2.7.2 Room Pressure Imbalance

This test identifies room pressure imbalances caused by closed doors in rooms with supply registers but no return registers.

1. Leave the house in winter conditions, and leave the air handler running.

2. Close interior doors.

3. Place hose from input tap on the manometer under one of the closed interior doors. Leave reference tap connected to outdoors.

4. Read and record this pressure measurement for each room. This pressure's magnitude indicates the degree to which the air handler's airflow is unbalanced between supply and return ducts in that room or zone.

If the pressure difference is more than ± 3.0 pascals with the air handler operating, pressure relief is necessary. To estimate the amount of pressure relief, slowly open door until pressure difference drops to between +3.0 pascals and –3.0 pascals. Estimate area of open door. This is the area required to provide pressure relief. Pressure relief may include undercutting the door or installing transfer grilles. For information on reducing duct-induced room pressures, *see page 64.*

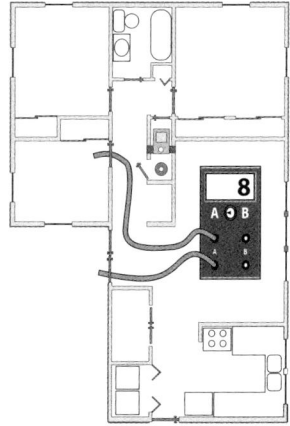

Blocked return path: With interior doors closed, the large positive pressure in the bedroom is caused by the lack of a air return register in the bedroom. The airflow in this forced air system is unbalanced, creating this pressure, and forcing room air through the room's air leaks to outdoors.

CHAPTER 3:   AIR SEALING HOMES

This chapter discusses the locations of air leaks and the methods and materials used to seal them. The job description of a building shell technician often includes duct sealing. Perform air leakage testing and evaluation before beginning air sealing or duct sealing work. *See "Diagnosing Shell & Duct Air Leakage" on page 23.*

3.1 AIR SEALING FUNDAMENTALS

Air leakage in homes accounts for 5% to 40% of annual heating and cooling costs. Air sealing is one of the most important energy-saving retrofits, and often the most difficult.

Air travels into and out of the building by three main pathways.

- ✓ Major air leaks, which are significant flaws in the home's air barrier
- ✓ Minor air leaks, which are often seams between building materials
- ✓ Through the building materials themselves (*See table 2-2 on page 40.*

Reducing air leakage accomplishes several goals.

- ✓ Saves energy by reducing unintentional air exchange with outdoors
- ✓ Saves energy by protecting the thermal resistance of the shell insulation
- ✓ Increases comfort by reducing drafts and moderating the radiant temperature of interior surfaces
- ✓ Reduces moisture migration into building cavities
- ✓ Reduces the pathways by which fire spreads through a building

3.1.1 Air Sealing Strategy

The first step to improving a building's airtightness is to formulate a strategy. Building a strategy starts by asking questions like these.

- ✓ What building components already serve as air barriers?

- ✓ How do we best seal the border areas between these components?

- ✓ Where there is no obvious air barrier, where do we establish one?

- ✓ How do we seal various penetrations through the air barrier?

- ✓ How does whole-house ventilation relate to shell air leakage? *See "Evaluating Home Ventilation Levels" on page 34.*

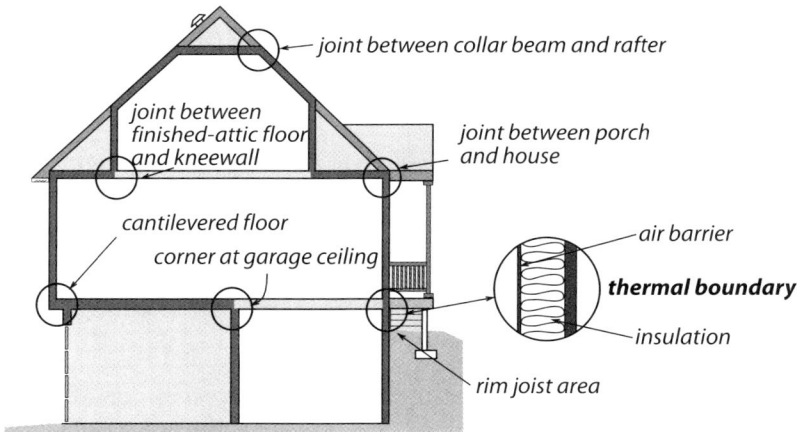

Thermal boundary flaws: The thermal boundary contains the air barrier and insulation, which should be adjacent to one other. The insulation and the air barrier are often discontinuous at corners and transitions.

3.2 Air Sealing Safety

Air sealing reduces the exchange of fresh air in the home, and can alter the pressures within the home. Before air sealing, survey the home to identify both air pollutants that may be concentrated by air sealing efforts and open combustion appliances that may be affected by changes in house pressure.

Don't perform air sealing when there are obvious threats to the occupants' health, the installers' health, or the building's durability that are related to air sealing. If any of the following circumstances are present, either postpone air sealing until they're corrected or correct the problems as part of the project's work scope.

- ✓ Measured carbon monoxide levels exceed the suggested action level. *See "Carbon Monoxide (CO) Testing" on page 208.*

- ✓ The building is below its Minimum Ventilation Requirement (MVR), and no mechanical ventilation exists or is planned. *See "Evaluating Home Ventilation Levels" on page 34.*

- ✓ Combustion zone depressurization exceeds -4 pascals during a worst-case test. *See "Worst-Case Depressurization, Spillage, and CO" on page 212.*

- ✓ Combustion appliance chimneys don't meet minimum standards.

- ✓ Unvented space heaters will be used after air sealing work.

- ✓ Moisture has caused structural damage or respiratory hazards from rot, mold, or dust mites. *See "Moisture Problems" on page 219.*

- ✓ Infestations, vermin, or other sanitary issues are present.

Install a whole-house ventilation system when performing effective home air sealing. *See "Other Pollutants" on page 224.*

3.3 AIR SEALING MATERIALS

Air barriers must be able to resist severe wind pressures. Use strong air barrier materials like plywood, drywall, galvanized steel, or foamboard to seal large air leaks, especially if your region has powerful winds. These strong materials should be attached with mechanical fasteners and construction adhesive.

Caulk should be only be used for sealing small cracks. Use liquid foam for cracks larger than $1/4$ inch.

3.3.1 Air Barrier Materials

Air barrier materials should themselves be air barriers. Perforated asphalt felt, concrete block, and earth are among the substantial materials that aren't considered air barriers. The materials discussed below are air barriers. *See also "Building Components and Their Air Permeance" on page 40.*

Plywood, OSB, etc.

Three-eighths-inch plywood and oriented strand board (OSB) are strong enough to resist any windstorm that spares the house. Although combustible and flammable, plywood, OSB, and masonite can be used as ignition barriers and fire stops. Attach these structural sheets with screws or nails and construction adhesive which strengthens the bond and accomplishes air sealing at the joint.

Drywall

Half-inch drywall constitutes a 15 minute thermal barrier, and is also an ignition barrier. When air sealing a fire-rated assembly in a commercial or multi-family building, choose drywall and a fire-rated caulking whenever possible. Fasten drywall with screws and construction adhesive. Don't use drywall where it will get wet or in damp locations.

Galvanized Steel

Being non-combustible, galvanized steel is used to seal around chimneys and other heat producing components. It can be used to build air seals around recessed light fixtures. A round galvanized duct with a galvanized cap makes a good air seal for round recessed light fixtures. To seal around chimneys, cut the galvanized steel accurately so that you can seal the gap with high temperature silicone or stove cement.

Foam Board

One- or two-inch foil-faced polyurethane foam board is easy to cut and fit and is an excellent insulating material. However, foam board is difficult to fasten sufficiently to resist a strong wind event. Plywood, OSB, drywall or other structural materials are far more suitable for sealing large air leaks because of the strong assembly that results by attaching structural air sealing materials to existing structural materials.

Cross-Linked Polyethylene House Wrap

House wrap is vapor permeable, waterproof, and a good air barrier. However it isn't rigid and can be pulled off its fasteners by powerful wind. When used as an air barrier underneath siding, it must be taped at the seams and protected from damage until siding is installed. House wrap is not a very good retrofit air sealing material.

Adhesive Window Flashing

Construction tape and adhesive window flashing can be very effective air sealing materials when used correctly with house wrap or vapor barrier material. Surface preparation is extremely important when using adhesive-backed materials because dust, oil, and moisture can cause the adhesive to fail. The flashing should lay very flat without any tenting in order to allow one-part foam to seal the gap between a new window and the rough opening.

3.3.2 Stuffing Materials

Stuffing materials are used to insulate a cavity, to give the cavity a bottom, or to serve as supporting part of an air seal.

Backer Rod

Backer rod is closed-cell polyethylene foam that creates a bottom barrier in a gap that will be caulked. Backer rod doesn't bond to the caulking, and so prevents three-sided adhesion that could tear the caulking bead apart with the expansion and contraction of temperature extremes.

Fiberglass Batts

Fiberglass batts are far from being an air barrier, even if stuffed tightly. However, they can be used to fill cavities for insulation, convection reduction, and support for two-part foam sprayed over the cavity.

Blown Cellulose

Blown cellulose is considered an air barrier because of its tendency to slow air migration by becoming wedged into gaps and cracks. However it is not much better than fiberglass at stopping air movement. When used to completely fill cavities, it can reduce blower door readings significantly. However, blown cellulose isn't strong enough to resist a windstorm which would blow a hole through the plug of cellulose or simply move it out of the way. Cellulose should never be used in areas where it may get wet.

3.3.3 Caulking and Adhesives

The performance of caulking and adhesives depends on their formulation and on the substrate to which they are applied. Some caulks and adhesives are quite sensitive to dirt and only work well on particular substrates, while others are versatile and dirt-tolerant. Removing debris and cleaning the joint are required for a long lasting seal.

Water Based Caulks

A wide variety of paintable caulks are sold under the description of acrylic latex and vinyl. These are the most commonly used caulks and the easiest to apply and clean up. Siliconized latex caulks are among the most adhesive and durable sealants in this group. Don't apply water based caulks to building exteriors when rain is expected since they are not waterproof until cured, and they stain nearby materials if they are rained upon while curing. Don't apply water based caulks when freezing weather is expected.

Silicone Caulk

Silicone has great flexibility, but its adhesion varies among different substrates. It is very easy to gun even in very cold weather. It's not as easy to clean up as water based caulks, though it's easier than polyurethane or butyl. Silicone is not paintable, so choose an appropriate color. High temperature silicone can tolerate temperatures above 400° F and is used with galvanized steel to air seal around chimneys.

Polyurethane Caulk

Polyurethane has the best adhesion and elasticity of any common caulk. It works very well for cracks between different materials like brick and wood. It resists abrasion and is used to seal critical joints in concrete slabs and walls. It is also good for sealing the fastening fins of windows to walls. It is almost as sticky and adhesive as a construction adhesive. Cleaning it up is difficult so neat workmanship is essential. Polyurethane caulk does not gun easily, and should be room temperature or higher. Polyurethane caulk doesn't hold paint.

Acoustical Sealant

This solvent based or water based adhesive is used to seal laps in polyethylene film and house wrap. Acoustical sealant is very sticky, adheres well to most construction materials, and remains flexible for years after application. Acoustical sealant is used to seal building assembles for sound deadening.

Water Soluble Duct Mastic

Duct mastic is the best material for sealing ducts, including cavities used for return ducts. A messy but highly effective sealant, duct mastic can be applied with a medium thickness brush or rubber glove. Have a bucket of warm water handy to clean your gloved hands and a rag to dry the gloves. Spread the mastic and use fiberglass fabric web tape to reinforce cracks more than $1/8$-inch in diameter. Thorough cleaning of dust and loose material isn't necessary. Mastic bonds tenaciously to everything, including skin and clothing.

Stove Cement

Used to seal wood stove chimneys and cement wood stove door gaskets in place. Withstands temperatures to 2000° F.

Fire-Rated Caulk

Some elastomeric caulks are designed specifically for use in fire-rated assemblies. They can withstand flame and temperatures to 2000° F. Use this type of sealant when sealing air leaks in fire-rated assemblies in multifamily and commercial buildings.

Fire-Rated Mortar

Used in conjunction with foam to seal various sized holes and gaps in commercial buildings with fire-rated building assemblies. This mortar covers the foam to preserve a non-combustible surface.

Construction Adhesives

Construction adhesives are designed primarily to bond materials together. But they also create an air seal if applied continuously around the perimeter of a patch. They are often used with fasteners like screws or nails but can also be used by themselves. Higher quality construction adhesives can be used as contact adhesives to bond lightweight materials. Apply the adhesive to one surface, touch the patch down to spread the glue to both surfaces, then remove the patch to expose the adhesive to air. After a few minutes, put the patch back in place. Make sure you put the patch in the correct location because it will be extremely difficult to remove.

Use specially designed construction adhesives for polystyrene foam insulation because a general purpose adhesives decompose the foam's surface.

3.3.4 Liquid Foam Air Sealant

Liquid closed-cell polyurethane foam is a versatile air sealing material. Closed-cell foam is packaged in a one-part injectable variety and a two-part sprayable variety. It has a very high R-value per inch and is idea for insulating and air sealing small, poorly insulated, and leaky areas in a single application.

Installation is easy compared to other materials to accomplish the same air sealing tasks. However, cleanup is difficult enough that you probably don't want to clean up multiple times on the same job. Instead identify all the spots needing foam application, make a list, and foam them one after another.

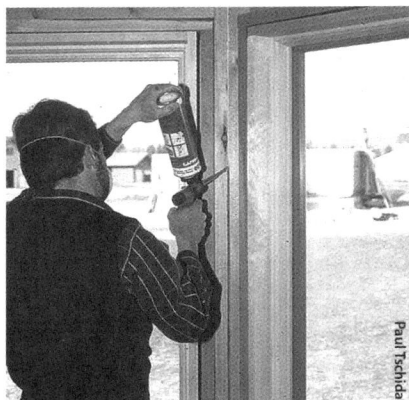

One-part foam: A technician uses an applicator gun to seal spaces between framing members and around windows.

Two-part foam: A technician air seals and insulates around an attic hatch dam with two-part spray foam.

One-Part Foam

This gap filler has tenacious adhesion. One-part foam is best applied with a foam gun rather than the disposable cans. Cleanup is difficult if you get careless. When squirted skillfully into gaps, this material reduces air leakage, thermal bridging, and air convection through the assemblies to which it is applied. One-part foam isn't effective or easy to apply to gaps over about one inch or to bottomless gaps. This product can leave small air leaks unless applied with skill.

Two-Part Foam

Good for bridging gaps larger than one inch. Two-part foam has become very popular for use with polyurethane foam board to sealing large openings. Cut foam board to close-enough tolerances around obstacles and fill the edges with the two-part foam. Two-part foam should be sprayed to at least an inch of thickness when it serves as an adhesive for foamboard patches over large holes for strength.

Caution: Wear a tight-fitting organic-vapor respirator when installing two-part foam and power-ventilate the work area.

Foam Construction Adhesive

Polyurethane foam dispensed from foam guns is an excellent adhesive for joining many kinds of building materials. It works well in joining foam sheets together into thick slabs for access doors through insulated building assemblies.

3.4 SEALING MAJOR AIR LEAKS AND BYPASSES

Major air sealing involves finding and sealing large openings that admit outdoor air into the conditioned space. Sealing major air leaks is one of the most cost-effective energy-saving measures, especially in cold climates.

Major air sealing activities are generally completed prior to insulating where insulation would be an obstacle to air sealing. Effective air sealing results in a significant drop in reduced blower door readings and/or changes in pressure diagnostics readings. *See "House Airtightness Testing" on page 26.*

Bypasses are pathways that allow air to migrate around the insulation. Bypasses are sometimes direct air leaks between indoors and outdoors and sometimes not. In many cases, bypasses merely allow air from indoors or outdoors to circulate around within the building cavities. They can still be a significant energy loss, however, when they are adjacent to interior surfaces of the home where circulating air can convect heat into or out of the home.

Major air leaks are often found between the conditioned space and intermediate zones such as floor cavities, attics, crawl spaces, attached garages, and porch roofs. The time and effort you spend to seal major air leaks should depend on its size.

Homes with finished attics

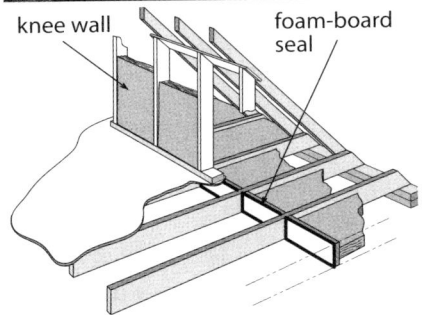

Sealing under knee walls: Here a n undersized piece of foam board is sealed into the joist spaces with one-part foam, forming a strong and airtight seal.

Major air leaks are not always easily accessible. When they are hard to reach, technicians sometimes blow dense-packed cellulose insulation into surrounding cavities, hoping that the cellulose will resist airflow and plug cracks between building materials.

3.4.1 Removing Insulation for Air Sealing

Attics are a particularly critical area for air sealing. The cost of removing insulation from an attic for the purpose of air sealing the attic may be well worth the price. Batts and blankets can be rolled up, moved out of the way, and re-used. Loose fill insulation can be vacuumed with commercial vacuum machines available from the same suppliers that sell insulation-blowing machines. Many insulation companies have these large vacuums already.

3.4.2 Major Air Leak Locations & Treatments

This section provides a list of the most important large air leakage areas commonly found in North American residential buildings. It provides general suggestions for air sealing these areas.

Joist Cavities Under Knee Walls

Floor joist cavities beneath knee walls allow air from a ventilated attic space to enter the floor cavity between stories. This is a problem of homes with a finished attic, also known as half-story homes.

Connect the knee wall with the plaster ceiling of the space below by creating a rigid seal under the knee wall. Use a combination of rigid foam with one-part or two-part foam sealing the perimeter. Or, use fiberglass batt or blown cellulose with spray two-part foam as a strong airtight seal covering over the cellulose or fiberglass batt.

Kitchen or Bathroom Interior Soffits

Many modern homes have soffited areas above kitchen cabinets and in bathrooms. Large rectangular passages link the attic with the soffit cavity. At best, the air convects heat into, or out of, the conditioned space. At worst, attic air infiltrates the conditioned space through openings in the soffit or associated framing.

Seal the top of the soffit with foam board, plywood, or drywall fastened and sealed to ceiling joists and soffit framing. Seal the patch's perimeter with two-part foam or caulking.

Anthony Cox

air leakage

eave vents

recessed light

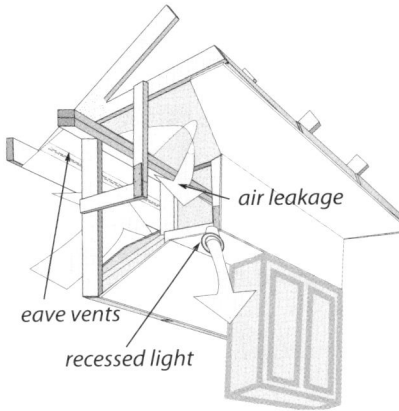

Kitchen soffits: The ventilated attic is connected to the soffit and the wall cavity through framing flaws. Any hole in the soffit creates a direct connection between the kitchen and attic. The photo shows a soffit sealed from the attic with foam board reinforced with two-part spray foam.

Plumbing Penetrations

Seal gaps with expanding foam or caulk. If the gap is too large, stuff it with fiberglass insulation, and spray foam over the top to seal the surface of the plug.

Seal holes and cracks from underneath with expanding foam. Seal large openings with rigid materials caulked or foamed at edges.

Sealing large plumbing penetrations: Foam board is attached with screws and washers. Gaps around the penetrations are filled with one-part foam to form a complete airtight seal.

Two-Level Attics in Split-Level or Tri-Level Houses

Split-level and tri-level homes have a particular ai leakage problem related to the walls and stairways dividing the homes' levels.

Seal wall cavities from the attic with a rigid material fastened to studs and wall material. Or insert folded fiberglass batt and spray with at least one inch of two-part foam to create a rigid air seal. Stapling house wrap to the exposed wall over the insulation retards both air leakage and convection.

Also seal all penetrations between both attics and conditioned areas.

Tri-level home

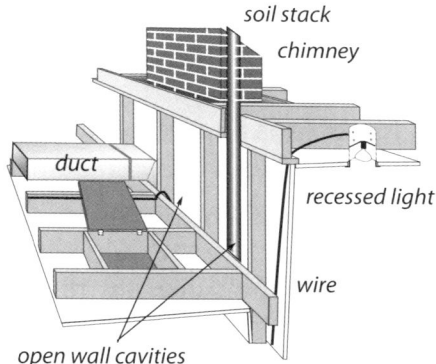

Two-level attic: Split-level homes create wall cavities connected to the ventilated attic. Other air leaks shown are duct, recessed light, and chimney.

Fireplace Chimneys

Fireplace chimneys are some of the most dramatic and serious air leaks commonly found in homes. The existing damper or "airtight" doors are seldom a good air seal. There are a number of commercial solutions for leaky or non-existent fireplace dampers. They include new tight dampers, tight top sealing dampers, and inflatable pillows. If the fireplace is never used, it can be air sealed from the roof with a watertight, airtight seal and from the living space with foamboard, covered with airtight drywall. If you install a permanent chimney seal such as this, be sure to install a notice in the fireplace warning future occupants of the blockage.

Spaces Around Chimneys

Seal chimney and fireplace bypasses with sheet metal (minimum 26 gauge thickness). Fasten the sheet metal with galvanized nails or screws and construction adhesive. Seal the metal patch to chimney or flue with a high temperature silicone sealant or chimney cement.

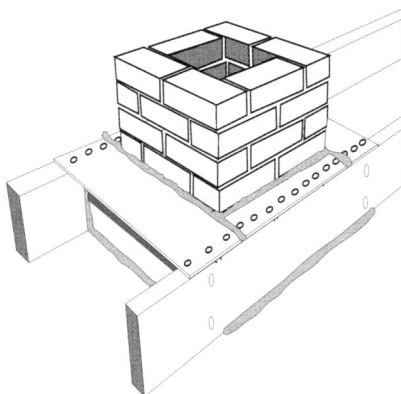

Sealing around chimneys: Chimneys require careful selection of materials for air sealing.

Exhaust Fans and Recessed Lights

Consider the following options for air sealing recessed light and fan fixtures and enhancing the fire safety of the ceiling assembly.

- ✓ Replace the existing recessed fixture with one rated: Insulation Contact Air Tight (ICAT).

- ✓ Caulk joints where housing comes in contact with the ceiling with high temperature silicone sealant to reduce air leakage around the housing.

Airtight IC recessed light fixtures: Use these fixtures to replace leaky existing fixtures.

- ✓ Build a metal (26-gauge galvanized steel) or drywall box over the fixture leaving at least 3 inches clearance from insulation on all sides including between the top of the fixture and the lid of the enclosure. Seal the airtight box to surrounding materials with foam to form an airtight assembly.

- ✓ Remove the recessed light fixture and patch the opening. Many homes have more light fixtures than necessary.

- ✓ Replace the recessed fixture with a surface mount fixture and carefully patch and air seal the hole.

Recessed light fixtures should contain 60-watt bulbs or less. Compact fluorescent lamps (CFLs) are preferred for recessed light fixtures because CFLs run cooler than incandescent lamps.

Caution: Don't cover IC-rated or airtight IC-rated fixtures with spray foam insulation. The high R-value and continuous contact could overheat the fixture.

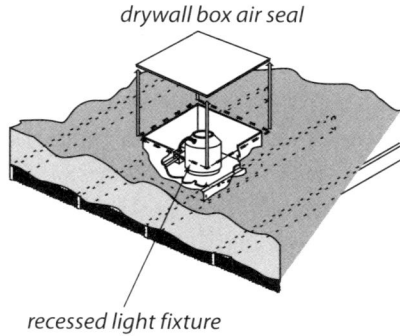

drywall box air seal

recessed light fixture

Recessed light fixtures: These are major leakage sites, but these fixtures require some air circulation to cool their incandescent bulbs.

Attic Hatches and Stairways

There are a wide variety of building assemblies for providing access from the building to an insulated attic. These access doors or panels and the framing and sheeting surrounding them often constitute a major air leakage path. Consider the following improvements.

- ✓ Weatherstrip around doors and hatches.

- ✓ Seal gaps around frame perimeters with one-part foam, two-part foam, or caulking.

- ✓ Clear fibrous insulation from around the hatch framing and spray two-part foam around the perimeter to reduce heat loss through the hatch framing.

Incomplete Finished Basements

Missing ceilings, wall segments, or rim joist sealing can allow heated basement air to circumvent the finished and insulated wall, carrying heat with it. Complete walls and ceiling or at least install air barriers between finished and unfinished areas. Seal edges of discontinuous walls thoroughly. Seal and insulate untreated rim joists.

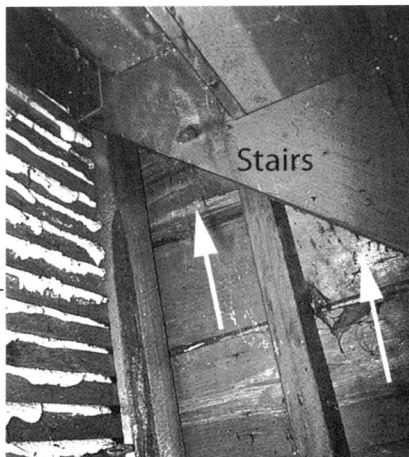

Leakage into attic: A closeted space beneath attic stairs provides large air leakage channels into the ventilated attic.

Porch Roof Structures

Porch roofs on older homes were often built at the framing stage or before the water resistive barrier (WRB) and siding were installed. The sheeting around the porch roof sheathing, roofing, and tongue-and-groove ceiling are not air barriers. The plank wall sheathing or unsheathed wall allows air into the wall cavities where it can migrate into the conditioned space or convect heat into or out of the building. This problem requires removing part of the porch ceiling or blowing the entire roof cavity full of cellulose insulation to reduce the airflow through the roof and wall cavities.

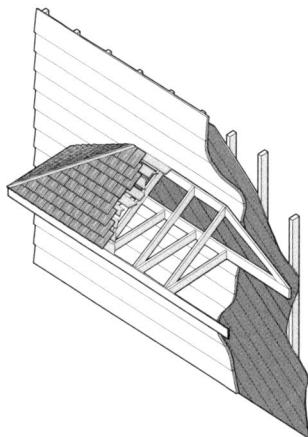

Porch air leakage: Porch roof cavities often allow substantial air leakage because of numerous joints, and because there may be no siding or sheathing installed in hidden areas.

Built-In Cabinets/Shelves

Built-in cabinets and shelves are a feature of older homes and present challenges for air sealing. Sealing these areas from inside the cabinet requires care and attention to aesthetics. Use caulking that is compatible with the colors of the surrounding wood and plaster. If possible, establish both an air barrier and insulation behind the cabinet, out of sight of the occupants.

Pocket Door Cavities

When located on the second floor, cap the top of the entire wall cavity in the attic with rigid board, caulked and mechanically fastened.

Cantilevered Floors

Floors that hang over their lower story are called cantilevered floors. The underside of the overhanging floor can leak considerably. If possible, remove a piece of soffit to determine the condition of insulation and air barrier. Use foam board, liquid foam, and caulking as necessary to create an air barrier at your chosen thermal boundary. Many balconies have cantilevered floors that leak air into a building's floor cavity.

Balloon Framed Walls

Balloon framed two-story walls are common in older homes, built when you could buy straight 20-foot 2-by-4s. Some modern homes have balloon framed gable walls, where the studs rise above the level of the ceiling joists and are cut at an angle to frame the gable. Even when these balloon framed gable walls are full of insulation, air can convect through the

Balloon framing at the second floor: Floor joists sit on a ledger board and are nailed to the stud. The floor cavity is open to the wall cavity and the framing cavities may be connected to the attic and basement.

insulation. On occasion, windstorms have actually blown the insulation out of the wall cavity into the attic.

The floor joists are often nailed to long wall studs, creating a continuous cavity through uninsulated wall into the uninsulated floor. Insulate the wall cavities to reduce air leakage and convection. Seal stud cavities from the attic, basement, or crawl space with a fiberglass insulation plug, covered with a 2-part foam air seal. Or seal the cavities with a rigid barrier, such as $1/4$-inch plywood or 1-inch foam board sealed to surrounding materials with caulk or liquid foam.

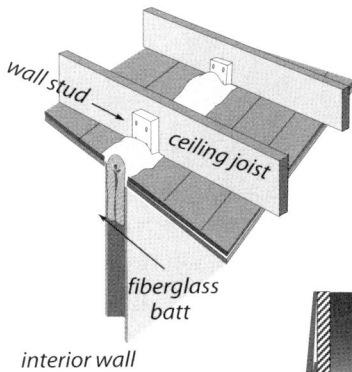

wall stud

ceiling joist

fiberglass batt

interior wall

Balloon framed interior walls: Fiberglass insulation covered by a 1-inch layer of two-part foam seals wall cavities.

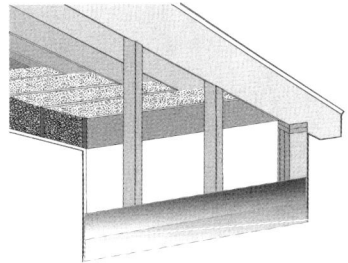

blown cellulose

Balloon framed gable: Studs extend above the ceiling allowing convection from the attic.

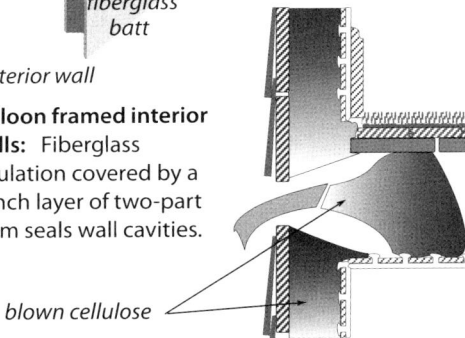

Sealing wall-floor junction: Blown cellulose reduces convection through walls and floors.

3.4.3 Sealing Duct Leaks

Ducts are often considered part of the thermal boundary because they are full of house air. Return ducts that are connected to attics, crawl spaces, or attached garages can create a lot of air leakage.

Ducts located outside the thermal boundary or in an intermediate zone like a ventilated attic or crawl space should be sealed. The following is a list of duct leak locations in order of their relative importance. Leaks nearer to the air handler are exposed to higher pressure and are more important than leaks further away.

Sealing Return Ducts

Return leaks are important for combustion safety and for efficiency, especially in hot humid weather. Consider using the following techniques according to how costly current duct leaks are.

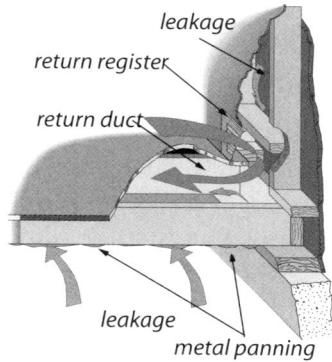

Panned floor joists: These return ducts are often very leaky and may require removing the panning to seal the cavity.

- ✓ First, seal all return leaks within the combustion zone to prevent this leakage from depressurizing the combustion zone and causing backdrafting.

- ✓ Seal panned return ducts using mastic to seal all cracks and gaps within the return duct and register.

Lining a panned cavity
Foil-faced foam board, designed for lining cavities is sealed with duct mastic to provide an airtight return.

- ✓ Seal leaky joints between building materials composing cavity return ducts, like panned floor cavities and furnace return platforms.
Remove the panning to seal cavities, containing joints in building materials.

- ✓ Carefully examine and seal leaks at transitions between panned floor joists and metal trunks that change the direction of the return ducts. You may need a mirror to find some of the biggest return duct leaks in these areas.

✓ Seal filter slots with an assembly that allows easy changing of filters.

✓ Seal the joint between the furnace and return plenum with silicone caulking or foil tape.

Sealing Supply Ducts

Inspect the following areas and seal them appropriately.

✓ Plenum joint at air handler: These joints may have been difficult to fasten and seal because of tight access. Seal these thoroughly even if it requires cutting an access hole in the plenum. Use silicone caulking or foil tape instead of mastic and fabric mesh here for future access – furnace replacement, for example.

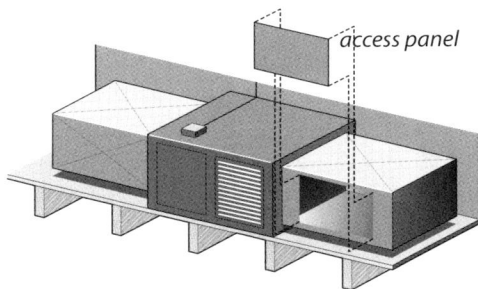

access panel

Plenums, poorly sealed to air handler: When air handlers are installed in tight spaces, plenums may be poorly fastened and sealed. Cutting a hole in the duct may be the only way to seal this important joint.

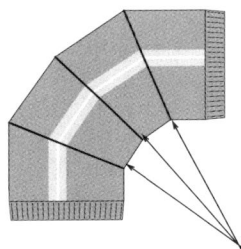

Sectioned elbows: Joints in sectioned elbows known as gores are usually quite leaky and require sealing with duct mastic.

✓ Joints at branch takeoffs: These important joints should be sealed with a thick layer of mastic. Fabric mesh tape is a plus for new installations or when access is easy.

✓ Joints in sectioned elbows: Known as gores, these are usually leaky and require sealing with duct mastic.

- ✓ Tabbed sleeves: Attach the sleeve to the main duct with 3-to-5 screws and apply mastic plentifully. Or better, remove the tabbed sleeve and replace it with a manufactured take-off.

- ✓ Flexduct-to-metal joints: Clamp the flexduct's inner liner with a plastic strap, using a strap tensioner. Clamp the insulation and outer liner with another strap. Clamping the inner liner tightly is sufficient for a tight seal without applying mastic.

- ✓ Deteriorating ductboard facing: Replace ductboard, preferably with metal ducting when the facing deteriorates because this condition leads to massive air leakage.

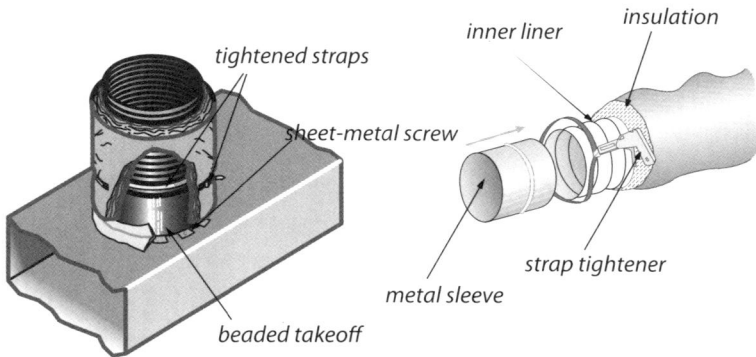

Flexduct joints: Flexduct itself is usually fairly airtight, but joints, sealed improperly with tape, can be very leaky. Use methods shown here to make flexduct joints airtight.

- ✓ Consider sealing off supply and return registers in unoccupied basements or crawl spaces.

- ✓ Seal penetrations made by wires or pipes traveling through ducts. Or better yet, move the pipes and wires and patch the holes.

Support ducts and duct joints with duct hangers approximately every 24 inches.

Duct Boots, Registers, and Chases

Caulk or foam the joint between the boot and the ceiling, wall, or floor between conditioned and unconditioned areas.

If chase opening is large, seal with a rigid barrier such as plywood or drywall, and seal the new barrier to ducts with caulk or foam. Smaller cracks between the barrier and surrounding materials may be foamed or caulked.

3.4.4 Minor Air Sealing

Minor air sealing includes sealing small openings with such materials as caulk, weather stripping, or sash locks. These measures tend to please the home's occupants by reducing perceived drafts, slowing the entry of dirt, or making the interior paint look better. But they rarely result in significant blower door reductions or changes in pressure diagnostic readings.

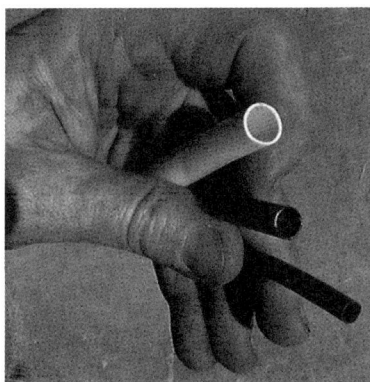

Silicone bulb weatherstrip: Silicone bulb has its own adhesive or is adhered to surfaces with silicone caulking.

Window and Door Frames

Sealing from the exterior serves to keep bulk water out and protect the building. If the crack is deeper than $5/16$-inch, it should be backed with a material such as backer rod and then sealed with caulk. Any existing loose or brittle material should be removed before the crack is recaulked.

Foundation Sill Plate

The rim joist area is composed of several joints. They can be sealed from the basement or crawl space with caulk or foam. Remove dust before applying sealant.

Masonry Surfaces

Best sealed with a cement-patching compound, mortar mix, or high quality caulking, such as polyurethane. For cement-based patches, consider priming the damaged areas with a concrete adhesive.

Leaky top plates: The cracks along top plates are from lumber shrinkage. They are small cracks but there are long lengths of them.

Interior Wall Top Plates

Drywall is installed after interior walls are constructed. The top plates of interior walls are open to the attic. Top plate shrinkage opens up cracks that run the entire length of the interior wall. Move insulation and seal the cracks with caulking or two-part foam.

Interior Joints

These can be caulked if blower door testing indicates substantial leakage. These joints include where baseboard, crown molding and/or casing meet the wall, ceiling, and/or floor surfaces. Gaps around surface mounted or recessed light fixtures and ventilation fans can be caulked if needed.

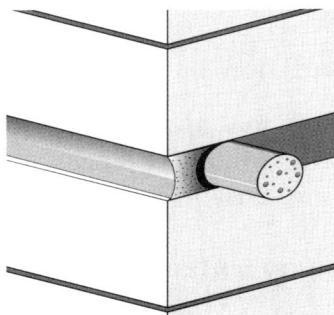

Backer rod: Use it to support caulk when sealing large uniform gaps. Use liquid foam for sealing irregular gaps.

3.4.5 Air Sealing Multi-Family Buildings

Large multi-family buildings present some unique air sealing issues to air sealing crews. Air moves in large older buildings through passages formed by construction features such as furring strips on exterior walls, duct chases, chimneys, stairways, plumbing stacks, and elevator shafts.

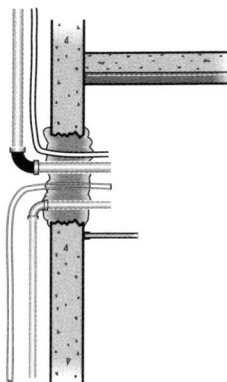

Large holes: Tradesmen often knock large holes in concrete walls without patching them. These can create large air leaks.

Experience has shown that large air leaks and bypasses must be sealed from the attic, not from inside the building, because there are just too many openings to seal once air enters walls, floors, and chases.

Follow these guidelines when air sealing older multi-family buildings.

✓ Don't rely on blower door testing. You may get false readings from air that enters through the other dwelling units.

✓ Focus on sealing large air leaks and bypasses that are found by visual inspection. Enter the attic space and seal openings with rigid barriers, foam sealant, and caulk.

✓ Perform moisture control measures to protect insulation. Don't install insulation if both bulk water and airborne moisture can't be controlled.

Air Sealing Homes

CHAPTER 4:   INSTALLING INSULATION

Insulation reduces heat transmission by slowing conduction, convection, and radiation through the building shell. Insulation combined with an air barrier creates the thermal boundary.

Installing insulation is one of the most effective energy saving measures. You can ensure its effectiveness by following these guidelines.

- ✓ Protect insulation from air movement with an effective air barrier. Make sure that the air barrier and insulation will be aligned (next to one another) using procedures outlined on *page 39*.

- ✓ Protect insulation from moisture by repairing roof and siding leaks, and by controlling vapor sources within the home. *See "Insulation Durability" on page 104*.

- ✓ Install insulation in a way that enhances fire safety and doesn't degrade it. *See "Fire Safety" on page 103*.

- ✓ Comply with lead-safe practices when disturbing paint in pre-1978 homes. *See "Lead-Safe Procedures" on page 231*.

- ✓ Install insulation to meet or exceed the guidelines of the International Energy Conservation Code (IECC) 2009. *See "Minimum R-Values from the IECC" on page 252*.

4.1 INSULATION MATERIAL CHARACTERISTICS

The purpose of insulation is to provide thermal resistance, which reduces the rate of heat transmission through building assemblies. Characteristics such as R-value per inch, density, fire safety, and installation effectiveness help technicians choose the right insulation for the job at hand. Insulation may have other useful qualities including vapor permeability or impermeability, and airflow resistance.

4.1.1 Fibrous Insulation Materials

Fibrous insulation materials are the most economical source of thermal resistance for buildings. If blown at a high density, fibrous insulations aren't air barriers but may contribute to the airflow resistance of an assembly that functions as an air barrier. The term *mineral wool*, as normally used, describes both fiberglass and *rock wool*. Rock wool is both a generic term and a trade name. We use rock wool in the generic sense as an insulating wool made from rocks or slag.

Cellulose was once made from virgin wood fiber under trade names like Balsam Wool. Now cellulose is manufactured from recycled paper and treated with boric acid or other fire retardants.

Fiberglass Batts and Blankets

Most fiberglass batts are either 14.5 inches wide or 22.5 inches wide to fit 16-inch or 24-inch wood stud or joist spacing. Batt R-values vary from 3.2 per inch to 3.7 per inch depending on density. Fiberglass blankets are available in a variety of thicknesses from 1 to 6 inches with vinyl and foil-skrim-kraft facings.

Batts must be installed very carefully to achieve their rated R-value. They insulate best when installed in contact with a good air barrier. Batts are packaged with a number of facings, including foil, kraft and vinyl, but unfaced batts are often the best choice and easiest to install. Although fiberglass is non-combustible, unfaced and faced batts contain flammable binder that holds the batt together. Fiberglass batts have flammable facings.

Fiberglass blankets are used to insulate metal buildings and to insulate crawl spaces from the inside. Although fiberglass doesn't absorb much moisture, the facings on blankets and batts can trap moisture which wets building materials and provides a water source for pests.

Blown Fiberglass

Loose fiberglass is blown in attics from 0.5 to 0.9 pcf and at that density the R-value is around 3.2 per inch. Blown fiberglass is non-combustible as a virgin product, but some blown fiberglass is made from chopped batt waste which contains a a small amount of flammable binder.

Fiberglass manufacturers now provide two blowing products, one for standard densities of up to 1.6 pcf, and another for dense-packing to more than 2 pcf.

In closed cavities, fiberglass is blown from 1.2 to 2.2 pcf, with R-value per inch from 3.6 to 4.2, increasing with higher density. The high density fiberglass is typically reserved for walls where the superior resistance to settling, airflow, and convection has extra value compared to lower density installations.

Blown Cellulose

Loose cellulose is blown in attics from 0.6 to 1.2 pcf and at that density range, the R-value is around 3.7 per inch. Expect around 15% settling within five years after installation.

In wall cavities, cellulose is blown at a higher density of between 3.5 to 4.0 pcf, to prevent settling and to maximize its airflow resistance. At the high density cellulose's R-value per inch is around 3.4. Evaluate the strength of wall cladding before blowing a wall to prevent damage during installation.

Cellulose is also mixed with sprayed water in damp-spray applications in open cavities and adhered to building surfaces. This cellulose material contains a non-corrosive fire retardant to prevent metal corrosion in the building assemblies where it is installed.

Cellulose is the most economical insulation and among the easiest to install. However, cellulose absorbs up to 130% of its own weight in water. Before you know of a moisture problem, the cellulose could be soaked, double its dry weight, and permanently degraded in thermal resistance. Because of its moisture absorption, avoid using cellulose in humid coastal regions or anywhere where it might encounter water or long periods of high relative humidity. We believe that cellulose shouldn't be installed in the following places regardless of climate: crawl spaces, floor cavities above crawl spaces or unconditioned basements, and horizontal or sloped closed roof cavities.

Rock Wool

Rock wool is a type of mineral wool like fiberglass. Rock wool has a small market share in North America. Rock wool batts have similar R-values per inch as fiberglass batts and contain flammable binders. Rock wool itself is non-combustible so blown rock wool will not burn.

4.1.2 Foam Insulation Materials

Foam insulation is commonly sprayed or injected as a liquid or as fastened to building assemblies as foam board. In general, foam insulations burn and create toxic smoke. Sometimes they must be covered by an ignition barrier of a thermal barrier when installed in attics and crawl spaces.

Foams are insect friendly materials that can aid termites in gaining a foothold in wood floor structures. All sources of ground water must be mitigated before applying foam to a foundation. When foam is installed on the outside of foundations, the surrounding soil should be treated with a termiticide. Inside a crawl space, foam should never provide a direct link from the ground to wood materials.

Closed-Cell Polyurethane Spray Foam

Polyurethane closed-cell foam, although the most expensive insulation discussed here, is a good value when space is limited, where an air vapor barrier is required, or where its structural strength and durability is needed. Polyurethane closed-cell foam is hazardous to installers and requires special personal protective equipment.

Polyurethane closed-cell foam typically installs at approximately 2 pcf density and achieves an R-value of 6 or more per inch. However, roofing applications call for a density near 3 pcf. Closed-cell foam is an air barrier and a vapor barrier.

Fire Protection for Closed-Cell Foam

Polyurethane closed-cell foam is combustible, sometimes flammable, and creates toxic smoke when it burns. When sprayed in attics and craw spaces, closed-cell foam must sometimes be covered by an ignition barrier or 15-minute thermal barrier.

Ignition barrier materials like galvanized steel and plywood are readily available and so is drywall, the most common 15-minute thermal barrier. However these materials aren't practical as coverings for most spray-foam applications. Intumescent paint, a new fire protection material, qualifies as either an ignition barrier or a thermal barrier, depending on the code jurisdiction.

Fire protection requirements vary among foam formulations, according to the amount and type of fire retardant. Foam manufacturers now have at least two formulations: normal and fire-resistant. The more fire-resistant type of closed-cell foam doesn't require an ignition barrier or a thermal barrier in most code jurisdictions.

Code jurisdictions and individual building officials vary in their interpretation of the IRC and other building codes with reference to plastic foams. Check with your local building officials before using closed-cell foam for fire-rated building assemblies.

Open-Cell Polyurethane Spray Foam and Injectable Foam

Polyurethane open-cell foam is installed at about 0.5 pcf and achieves an R-value of around 3.7 per inch. Newer formulations are available in densities of around 1.0 pcf with an R-value of around 4.7 per inch. Open-cell foam is available in a highly expansive spray formulation and a less expansive injectable formulation.

The formulation is injected through a hole, one inch or smaller, through an injection nozzle and not a fill tube. (The plastic fill tube would clog and is not cleanable.) The closed-cell foam can subject a wall cavity to a lot of pressure, so evaluate wall-cladding strength before injecting it.

Polyurethane open-cell foam has little structural strength. It can be trimmed to eliminate protruding excess from open wall cavities. Polyurethane open-cell foam is not a vapor barrier and is more porous to airflow than the closed-cell foam. It is difficult to install in deep cavities without creating voids.

Open-cell foam can absorb water vapor and liquid water. It can become a medium for mold growth. Don't use open-cell foam in crawl spaces unless proper drainage and ventilation have prevented ground moisture from being a problem.

Polyurethane open-cell foam is less flammable than similar closed-cell formulations but still creates toxic smoke when it burns and is a hazard to installers.

Air Krete Injectable Foam

Air Krete is a proprietary, low-expansion, non-toxic, and non-combustible insulation installed at around 1 pcf with an R-value of about 4.0 per inch. It is used when non-combustible insulation is required. It is water soluble and can be injected through a fill tube.

Tripolymer Injectable Foam

Tripolymer is a proprietary, low-expansion injectable foam installed at around 1.2 pcf at an R-value of 5.1 per inch. The material has very low flammability and smoke generation during a fire. It is widely used to insulate hollow masonry walls. It is water soluble and can be injected through a fill tube.

Expanded Polystyrene (EPS) Foam Board

EPS foam board, sometimes called beadboard, is the most economical of the foam insulations. EPS varies in density from 1 to 2 pcf with R-values per inch of 3.9 to 4.7, increasing with increasing density. EPS is packaged in a wide variety of products by local manufacturers, including structural insulated panels (SIPS), tapered flat-roof insulation, EPS bonded to drywall, and EPS embedded with fastening strips.

EPS is flammable and produces toxic smoke when burned. It has a low maximum operating temperature (160 degrees F) that is a concern for using EPS under dark-colored roofing or siding.

EPS is very moisture resistant and its vapor permeability is in the same order of magnitude as concrete, which makes EPS a good insulation for masonry walls.

Dense EPS (2 pcf) is appropriate for use below grade with weatherproof coverings to prevent degradation by ultraviolet light and freezing and thawing at ground level.

Extruded Polystyrene Foam (XPS) Board

XPS is produced by only a few manufacturers and is popular for below-grade applications. XPS is more expensive than EPS and has an R-value of 5.0 per inch. XPS may be the most moisture-resistant of the foam boards.

XPS is flammable and produces toxic smoke when burned. XPS must be covered by a thermal barrier when installed in living spaces. It has a low operating temperature (160 degrees F) that is a concern for using EPS under shingles or dark-colored siding. XPS must be protected from UV radiation and freezing and thawing at ground level.

Polyisocyanurate (PIC) Foam Board

PIC board has the highest R-value per inch (R-6 to R-7) of any common foam board. PIC is packaged with a vapor permeable or foil (vapor impermeable) facing. PIC is expensive but worth the cost when the depth of an insulation installation is limited.

PIC is combustible and produces toxic smoke during a fire. However some products have fire retardants that allow installation in attics and crawl spaces without ignition barriers. PIC has a low maximum operating temperature (<200 degrees F) that is a concern for using PIC under dark-colored roofing or siding.

Polystyrene Beads

Polystyrene (EPS) beads can be poured or blown into cavities. The cavities must be airtight or the beads will escape, making an annoying mess. EPS beads have an R-value between 2.2 and 2.5 per inch. They work well for filling hollow masonry walls.

Vermiculite and Perlite

These expanded minerals are pourable and used when a non-combustible insulation or high temperature insulation is needed. R-value per inch is between 2.0 and 2.7 per inch. These products are good for insulation around single-wall chimney liners to prevent condensation in the liner.

4.2 INSULATION SAFETY AND DURABILITY

Insulation activities require awareness about safety. Reference the following safety-related sections of this guide as appropriate.

- ✓ See *"Moisture Problems" on page 219.*
- ✓ See *"Worker Health and Safety" on page 235.*
- ✓ See *"Lead-Safe Procedures" on page 231.*

The following fire-safety and durability issues are particularly important to installing insulation.

4.2.1 Fire Safety

Consider a home's fire safety when air sealing, insulating, and all other related energy measures and repairs.

✓ Install non-insulating insulation shields around heat-producing devices like recessed light fixtures (not labeled IC). Fasten the shields securely to the ceiling so they don't collapse. Maintain 3 inches of clearance between the insulation and the fixture on the sides and top.

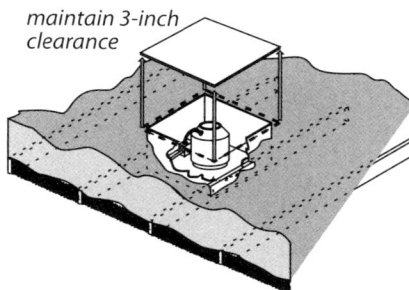

maintain 3-inch clearance

Recessed light fixtures: Covering recessed light fixtures with fire-resistant drywall or sheet-metal enclosures reduces air leakage and allows installers to safely insulate around the box.

✓ Use only non-combustible insulation to cover the top of insulation shields surrounding non-IC-rated fixtures.

✓ Fibrous insulation may surround and contact IC-rated fixtures (IC = insulation contact). However, spray foam insulation must not surround and contact IC-rated fixtures or any heat-producing fixtures.

✓ Install non-combustible insulation shields around masonry chimneys, B-vent chimneys, L-vent chimneys, and all-fuel chimneys to keep insulation at least 3 inches away from these chimneys.

✓ A non-combustible insulation or the listing of the vent material may allow either a smaller clearance than 3 inches or no clearance between insulation and the previously mentioned types of chimneys.

✓ Foam insulation requires a thermal barrier covering of at least half-inch drywall when installed in a living space. Foam may also require a thermal barrier or ignition barrier when installed in attics or crawl spaces. A *thermal barrier* is a material that protects combustible materials behind it from heat during a fire. An *ignition barrier* is designed to delay the ignition of the material it protects. Ignition barriers include plywood, galvanized steel, damp-spray fiberglass, and intumescent paint. Intumescent paint is a proprietary latex coating designed to protect foam insulation from flames and heat.

4.2.2 Insulation Durability

Moisture is the most common and severe durability problem in insulated building assemblies. Moisture fosters rot by insects and microbes. Entrained moisture reduces the thermal resistance of many insulation materials. Moisture affects the chemistry of some building materials.

Moisture prevention includes denying moisture access to building cavities, allowing condensed water to drain out, and allowing moisture to dry to the indoors, outdoors, or both.

Retrofitting insulation can affect the preventive measures listed here. Consider the function and relevance of these building components whenever you install insulation.

✓ **Air barrier:** Air can carry moisture into building cavities from indoors or outdoors where the moisture can condense and dampen insulation and other building materials. Air leakage is an energy problem too. The air barrier is any continuous material or building assembly that provides acceptable resistance to Air Leakage.

✓ **Vapor barrier:** Vapor diffusion can carry large amounts of water vapor into building cavities where it can condense and dampen insulation and other building materials. Vapor barriers prevent water vapor from indoors from diffusing into cavities where condensation can wet insulation and other building materials in cold climates. Cold climates have large differences in humidity between indoors and outdoors that can push vapor through building cavities.

✓ **Ground-moisture barrier:** The ground under a building is the most potent source of moisture in many buildings, especially those built on crawl spaces. Most crawl spaces require ground-moisture barriers to prevent the ground from being a major cause of moisture problems.

✓ **Water resistive barrier (WRB):** Asphalt paper or house wrap, under siding and roofing, serves as the home's last defense to wind-driven rain, which can dampen sheathing and other building materials. This water resistive barrier must be protected during insulation and incorporated into window openings during window replacement.

✓ **Vapor permeable materials:** Most common building materials are permeable to water vapor, which allows the water vapor to follow a gradient from wet to dry. This process allows building assembles to dry out to either the indoors and outdoors and is the hallmark of fail-safe building assemblies in most climates.

✓ **Flashings:** Seams and penetrations in building assemblies are protected by flashings, which prevent water from entering these vulnerable areas.

✓ **Drainage features:** Intentional or unintentional drainage features of buildings allow water to drain out of cavities. Examples: Masonry veneers have intentional drainage planes and weep openings near their bottoms. Cathedral ceilings drain water out through their soffit vents unintentionally.

✓ **Water storage:** Masonry veneers and structural masonry walls have the ability to store rainwater and dry out during dry weather.

✓ **Ventilation:** Roofs, attics, crawl spaces and even some walls have ventilation features that dry out wet building assemblies.

- ✓ **Termiticide:** When foam insulation is installed below grade in regions with termites, apply a termiticide to the soil in amounts determined by the labeling of the termiticide.

Consult with experts when necessary to preserve, protect, or install these moisture-prevention features, as appropriate according to local climate and established best practices.

4.3 ATTIC AND ROOF INSULATION

Attic insulation is one of the most cost-effective energy conservation measures available.

4.3.1 Preparing for Attic Insulation

Perform these preparatory steps before installing attic insulation.

- ✓ Before insulating the attic, seal air leaks and bypasses as described previously. Air leakage and convection can significantly degrade the thermal resistance of attic insulation. If attic air leaks are not properly sealed, increasing attic ventilation may increase the home's Air Leakage rate. *See "Major Air Leak Locations & Treatments" on page 79.*

- ✓ Repair roof leaks and remove other large moisture sources and repair other attic-related moisture problems before insulating attic. If attic-related moisture problems can't be repaired, don't insulate the attic.

- ✓ Vent all kitchen and bath fans outdoors through appropriate roof fittings, side wall fittings, or soffit fittings. Use galvanized steel vent pipe, and insulate the pipe to prevent condensation. Avoid using flexible plastic or aluminum duct because these materials restrict airflow. Check all fans for proper back-draft damper operation. Repair or replace the damper or the entire fan assembly if the damper doesn't operate freely.

- ✓ Install an attic access hatch if none is present, preferably at the exterior of the home. The attic hatch should be at least 22 inches on each side if possible. Insulate the hatch to the maximum practical R-value.

- ✓ Build an insulation dam around the attic access hatch. Build the dam with rigid materials like plywood or oriented-strand board so that the dam supports the weight of a person entering or leaving the attic.

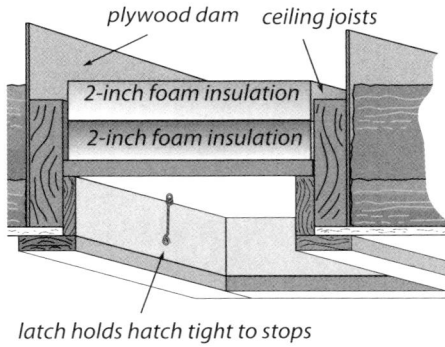

plywood dam   *ceiling joists*

2-inch foam insulation

2-inch foam insulation

latch holds hatch tight to stops

Insulated attic hatch: Building a dam prevents loose-fill insulation from falling down the hatchway. Foam insulation prevents the access hatch from being a thermal weakness. Install foam to achieve attic-insulation R-value or at least R-30. Foam can be glued together in layers.

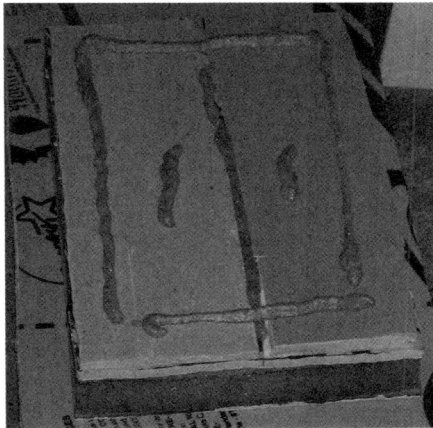

✓ Install vent chutes, baffles, tubes, or other devices to prevent blown insulation from plugging air channels between soffit vents and the attic. These devices should maximize the amount of insulation that may be installed over top plates without clogging ventilation paths. They also help prevent the wind-washing of insulation caused by cold air entering soffit vents.

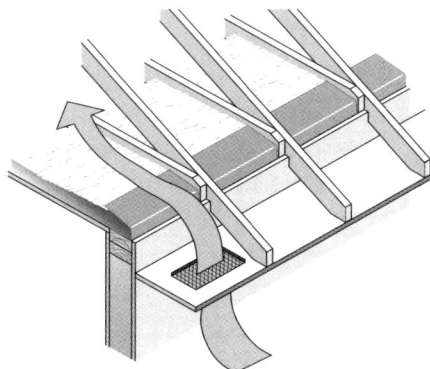

Soffit chute or dam: Allows installation of maximum amount of insulation in this cold area. Also prevents wind washing and airway blockage by blown insulation.

4.3.2 Attic Ventilation

Attic ventilation is intended to remove moisture from the attic during the heating season and/or to remove solar heat from the attic during the cooling season. Evaluate attic ventilation according to the 2009 International Residential Code (IRC) which states the following.

✓ A maximum ratio of one square foot of net free area to 150 square feet of attic area.

✓ With an interior vapor barrier or with distributed ventilation (high and low), only one square foot of vent per 300 square feet of attic area is required.

Retrofitting additional attic ventilation won't cure a moisture problem caused by airborne moisture migrating up from the living space. Instead, preventing moisture from entering the attic is the best way to keep attic insulation dry. Ceilings should be thoroughly air sealed to prevent moist indoor air from leaking through the ceiling.

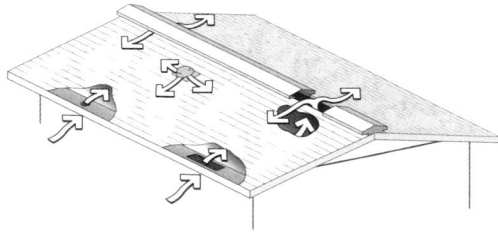

Low and high attic ventilation: Distributed ventilation — high and low — is more effective than vents that aren't distributed.

Power ventilators have limited value in reducing air conditioning cost and can consume a lot of electricity themselves. Many of these fans run much longer than they are needed, counteracting any benefit they may provide.

Unventilated Attics

According to the IRC 2009, attics may be unventilated if there is no vapor barrier on the ceiling and if the roof assembly is insulated with an air-impermeable insulation, like high-density polyurethane.

4.3.3 Blowing Attic Insulation

Install attic insulation to a cost effective R-value, depending upon existing insulation level and climatic region. air sealing attics must precede attic insulation and this may require removing existing insulation and debris that currently prevents effective air sealing. *See "Removing Insulation for Air Sealing" on page 78.*

Blown insulation is usually preferable to batt insulation because blown insulation forms a seamless blanket. Attic insulation always settles: cellulose usually settles 10% to 20% and fiberglass settles 3% to 10%. Blowing attic insulation at the highest achievable density helps minimize settling while minimizing air moment within the insulation.

Follow these specifications when installing loose-fill attic insulation.

- ✓ Calculate how many bags of insulation are needed to achieve the R-value specified on the bag's label. *See "Calculating Attic Loose-Fill Insulation" on page 253.*

- ✓ Maintain a high density by moving as much insulation as possible through the hose with the available air pressure. The more the insulation is packed together in the blowing hose, the greater is the insulation's density.

- ✓ When filling a tight eaves space, snake the hose out to the edge of the ceiling. Allow the insulation to fill and pack before pulling the hose back towards you. The eave area is the home's largest thermal bridge and it's important to put as much insulation into the eaves as possible, even if a little insulation spills into the soffit. The baffles provide the air space for ventilation.

- ✓ Fill the edges of the attic first, near the eaves or gable end, and move toward the center.

✓ Install insulation at a consistent depth. Level the insulation with a stick if necessary.

✓ Install insulation depth rulers: one for every 300 square feet.

✓ Post an insulation certificate near the attic entrance to facilitate inspection.

Blown-in attic insulation: Blown insulation is more continuous than batts and produces better coverage. Insulation should be blown at a high density to minimize settling and air convection.

4.3.4 Insulating Closed Roof Cavities

Many homes have cathedral ceilings or flat roofs that are only partially filled with insulation. The IRC 2009 building code requires a ventilated space above the roof insulation.

Many cathedral roof cavities have been dense-packed with fiberglass insulation without space for ventilation and some experts believe this is an effective solution. However, this solution requires engineered plans. Dense-packing a roof cavity with cellulose isn't a good option because of its high moisture absorption.

Incorporating Roof Ventilation

To prepare for roof-cavity insulation, incorporating a ventilated space above the insulation, observe the following steps.

✓ Demolish either the roof sheathing and roofing or the interior ceiling to gain full access to the cavity.

✓ Remove recessed light fixtures and replace them with surface-mounted ones. Carefully patch and air seal the openings.

✓ Install fiberglass or foam insulation to meet the IECC 2009 regional minimum roof-assembly R-value requirements. *See "Minimum R-Values from the IECC" on page 252.*

✓ Install openings into the ventilation channel above the insulation totalling $^1/_{150}$ of the roof area. If the ceiling has a vapor barrier the requirement becomes $^1/_{300}$ of the roof area.

✓ In cold climates, install a vapor barrier at the ceiling, at minimum an oil-based primer over the interior drywall or plaster.

✓ Repair roof leaks or install a new water-tight roof. Replace moisture-damaged sheathing as part of the roof replacement.

✓ Install an air-barrier ceiling (drywall) if the existing ceiling isn't an adequate air barrier, for example tongue-and-groove paneling.

✓ Seal other air leaks with great care, especially at the perimeter and abound ridge beams.

✓ When replacing the roof, consider installing 2-to-12 inches of rigid foam insulation on top of the roof deck. If you choose this option, dense-pack the existing roof cavity as described next.

Dense-Packing Closed Roof Cavities

Many homes have cathedral ceilings or flat roofs that are only partially filled with insulation. Whether these cavities can be dense-packed with fiberglass insulation is controversial and depends on climate. Dense-packing the cavities prevents most convection and infiltration, which are leading causes of moisture problems in roof cavities. Consult a knowledgeable local engineer before deciding to dense-pack a roof cavity with fiberglass. Cellulose isn't a good insulation choice for this application because of its high moisture absorption.

To prepare for roof-cavity insulation, observe the following steps.

- ✓ In cold climates, the ceiling should have a vapor barrier on the inside.

- ✓ Remove recessed light fixtures and replace them with surface-mounted ones. Carefully patch and air seal the openings.

- ✓ Seal other air leaks with great care.

- ✓ Reduce or eliminate sources of moisture in the home. *See "Preventing Moisture Problems" on page 222.*

- ✓ Don't insulate the roof deck without first filling the closed roof cavity because the roof assembly's thermal resistance will be reduced by convection inside the roof cavity.

Always use a fill tube when blowing closed roof cavities. The tube should reach into the cavity to within a foot of the end points of the cavity. Access

Blowing from the roof deck: Technicians remove a row of shingles, drill, and blow fiberglass into this vaulted roof cavity.

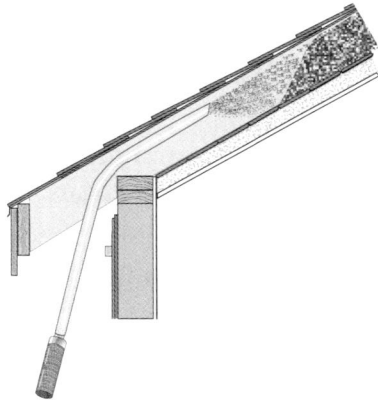

Blowing from the eaves: Some vaulted ceilings can be blown from the eaves and/or the ridge.

the cavity through the eaves, the roof ridge, the roof deck, or the ceiling. Consider the following procedures.

- ✓ Drilling holes in the roof deck after removing shingles or ridge roofing.
- ✓ Removing soffit and installing insulation from the eaves.
- ✓ Drilling through a drywall ceiling.
- ✓ Carefully removing a tongue-and-groove plank and filling cavities through this slot.

4.3.5 Installing Attic Batt Insulation

Follow these specifications when installing fiberglass batts in an attic. Fiberglass batts aren't the best insulation for attics because of all the resulting seams.

- ✓ Install unfaced fiberglass insulation whenever possible. Faced insulation doesn't tend to lay as flat as un-faced batts, and the facing is not very effective in slowing vapor movement because most of the vapor movement happens with air migration.
- ✓ If you must install faced batts, install them with the facing toward the heated space. Never install faced insulation over existing insulation.
- ✓ Cut batts carefully to ensure a tight fit against the ceiling joists and other framing.

4.3.6 Cathedralized Attics

Sometimes builders or retrofitters choose to insulate the bottom of the roof deck instead of the ceiling. Air handlers, installed in the attic, usually suggest this solution; to bring the air handler within the thermal boundary.

Insulating the roof deck presents some risk of moisture problems from vapor condensing in the insulation or at the roof deck. Most climates , especially cold and temperate climates, require air-impermeable insulation to avoid moisture condensation within the insulation or at the bottom of the roof deck. The methods below have been used to insulate roof decks successfully and are listed from the most moisture resistant to least moisture resistant.

✓ Four (4) or more inches of sprayed closed-cell polyurethane foam.

✓ Less than four (4) inches of sprayed closed-cell polyurethane foam with or without a layer of fibrous insulation below the foam.

✓ Six (6) or more inches of open-cell polyurethane foam.

✓ Six (6) or more inches of damp-sprayed fiberglass or cellulose insulation.

With any of these options, remove any existing vapor barrier from the ceiling assembly before insulating the roof.

If netting or fabric is used to contain fibrous insulation, the insulation should be blown with no voids at the highest achievable density allowed by the netting or fabric.

Some building departments require an ignition barrier of 1.5 inches of fibrous insulation or a proprietary liquid-applied coating to delay the spray foam's ignition during a fire.

4.3.7 Vaulted Attics

A vaulted attic is framed with a special truss that creates a sloping roof and a sloping ceiling. Access to the cavity varies from difficult to impossible. Install insulation either at the top of the roof deck or at the ceiling. Insulation, installed at the ceiling, must have some stability to prevent gravity from pulling it downhill or wind from piling it, leaving some areas under-insulated. Consider the following options to insulating uninsulated or partially insulated vaulted attics.

✓ Insulate the ceiling with fiberglass batts. Install the batts parallel to the framing if the top of existing insulation is below the framing. Install the batts perpendicular to the framing if the top of the existing insulation is above the framing.

✓ Insulate the bottom of the roof deck, as described previously for a cathedralized attic if the ceiling is removed or for new construction.

✓ Insulate the ceiling with sprayed foam, damp-spray fibrous insulation, or batts from the roof with the roof sheathing removed.

✓ Fill the cavity to approximately 90% with loosely blown fiberglass or cellulose from indoors or through the roof. Settling allows room for ventilation and air circulation within and around the insulation. The volume of insulation prevents major movement due to wind or gravity.

✓ Preserve or install openings into the ventilation space above the insulation totalling $^1/_{150}$ of the roof area. If the ceiling has a vapor barrier the requirement becomes $^1/_{300}$ of the roof area.

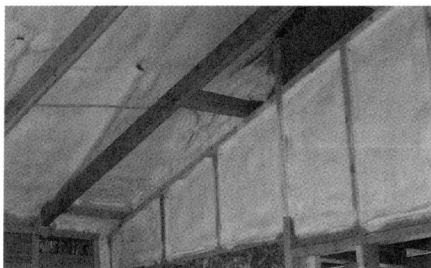

Foam insulated roof of vault: Vaults can be filled with stabilized insulation or their roof decks can be sprayed with foam.

4.3.8 Finished Knee Wall Attics

Finished attics require special care when installing insulation. They often include five separate sections that require different sealing and insulating methods. Seal air leaks in all these assemblies before insulating them. If necessary, remove the planking and insulation from the side-attic floor to expose the air leaks

Follow these specifications when insulating finished attics.

✓ Seal large air leaks between conditioned and non-conditioned spaces. *See "Air Sealing Homes" on page 67.*

✓ Inspect the structure to confirm that it has the strength to support the weight of the insulation.

✓ Insulate access hatches to the approximate R-value of the assembly through which it is located.

Exterior Walls of Finished Attic

Insulate these walls as described in *"Wall Insulation" on page 121.*

Collar-Beam Attic

Insulate this type of half-story attic as described in *"Perform these preparatory steps before installing attic insulation." on page 106.*

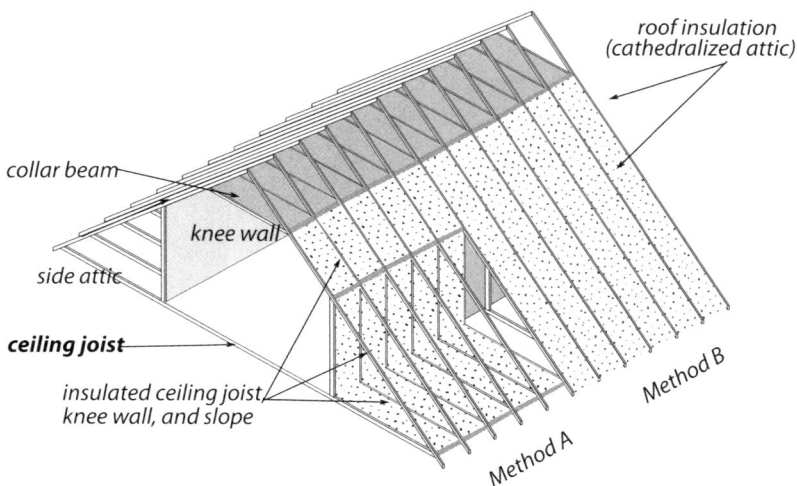

Finished attic: This illustration depicts two approaches to insulating a finished attic. Either A) insulate the knee wall and side attic floor, or B) insulate the roof deck.

Sloped Roof

Insulate sloped roof with densepack fiberglass or cellulose insulation. Install plugs of fiberglass batt, or other vapor permeable material, at the ends of this cavity to contain the blown insulation while allowing it to breathe.

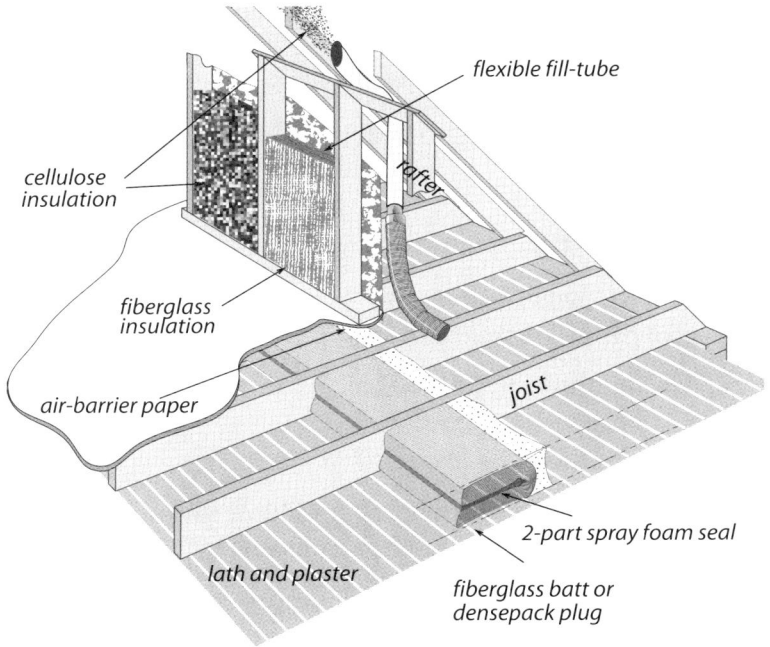

Finished attic best practices: Air sealing and insulation combine to dramatically reduce heat transmission and air leakage in homes with finished attics.

Side Attic: Outer Ceiling Joists

Insulate this small attic as described in *"Perform these preparatory steps before installing attic insulation." on page 106*. If this attic has a floor, remove one or more pieces of flooring to access the cavity and blow fiberglass or cellulose insulation to a medium pack.

Knee Walls

Insulate knee walls using any of the following methods.

- ✓ Install un-faced fiberglass batts and cover the insulation with house wrap on the attic side. Prefer R-15, 3.5-inch batts to the older lower-resistance batts.

- ✓ Install the house wrap first and blow fibrous insulation into the cavity through the house wrap.

- ✓ Spray the cavities with open-cell or closed-cell polyure-thane foam.

To seal and insulate under the knee wall, create an airtight and structurally strong seal in the joist space under the knee wall. This can be done by inserting 2-inch-thick foam sheets and foaming their perimeters with one-part or two-part foam, or by inserting a fiberglass batt into the cavity and foaming its face with an inch or two-part closed-cell spray foam.

For kneewall hatches, observe the following.

✓ Insulate knee-wall access hatches and collar-beam access hatch with 3 or more inches of rigid-foam insulation. Or install a fiberglass batt or batts wrapped with house wrap stapled to the hatch door.

✓ Weatherstrip the hatch and provide a positive closure.

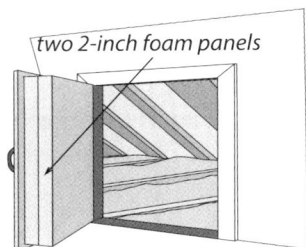

two 2-inch foam panels

Insulated access door in knee wall: Try to achieve at least R-15, or the highest R-value practical.

4.3.9 Walk-Up Stairways and Doors

Think carefully about how to establish a continuous insulation and air barrier around or over the top of an attic stairway. If the attic is accessed by a stairwell and standard vertical door, you can blow dense pack fibrous insulation into walls of the stairwell. Install a threshold or door sweep, and weatherstrip the door. Also, blow packed cellulose insulation into the cavity beneath the stair treads and risers.

You can also establish the thermal boundary at the ceiling level, but this requires a horizontal hatch at the top of the stairs.

When planning to insulate stairwells, investigate barriers such as fire blocking that might prevent insulation from filling cavities you want to fill, and consider what passageways may lead to other areas you don't want to fill such as closets. Balloon-framed walls and deep stair cavities complicate this measure.

Insulating & Sealing Retractable Attic Stairways

Retractable attic stairways are sometimes installed above the access hatch. Building an insulated box or buying a manufactured stair-and-hatchway cover are good solutions to insulating and sealing this weak point in the thermal boundary.

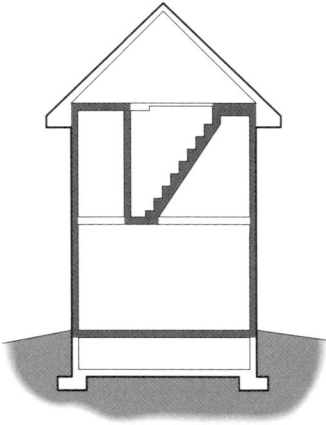

Insulating and sealing attic stair walls, doors, and stairs: Insulating and air sealing these is one way of establishing the thermal boundary.

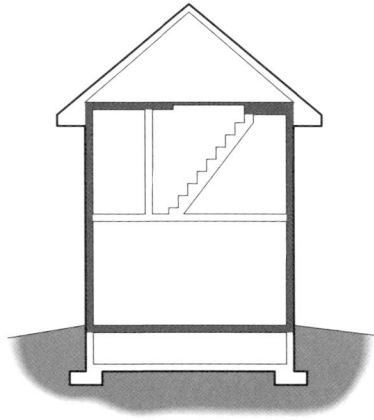

Insulating and weatherstripping the attic hatch: Air sealing around the hatch is an alternative way of establishing the thermal boundary.

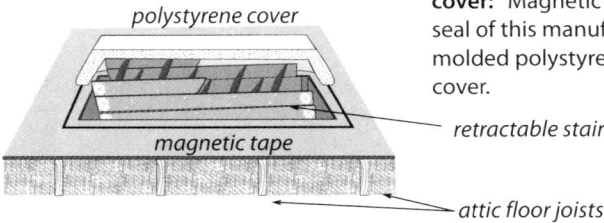

polystyrene cover

magnetic tape

retractable stair

attic floor joists

Manufactured retractable-stair cover: Magnetic tape forms the seal of this manufactured molded polystyrene insulated cover.

4.4 WALL INSULATION

If you find the existing walls uninsulated or partially insulated, add insulation to provide complete coverage for all the home's exterior walls.

Install wall-cavity insulation with a uniform coverage and density. Wall cavities encourage airflow like chimneys. Convection currents or air leakage can significantly reduce wall insulation's thermal performance if channels remain for air to migrate or convect.

In general, insulation specialists over-estimate the R-value of walls. Thermal bridging at studs, plates, and window and door framing along with poor installation can take up to 50 percent off the nominal value of the wall-cavity insulation.

The thermal bridging problem, in particular, may require installing foam sheathing under existing or new siding to cope with future energy costs.

Blown Wall-Insulation Types

Cellulose, fiberglass, and open-cell polyurethane foam are the leading insulation products for retrofit-installation into walls.

Table 4-1: Wall Insulation Density and R-Value per Inch

Insulation Material	Density	R-Value/in.
Fiberglass (virgin fiber)	2.2 pcf	4.1
Cellulose	3.5 pcf	3.4
Open-cell urethane foam	0.5 pcf	3.8
Tripolymer	1.2 pcf	5.1
Air Krete	2.1 pcf	3.9
pcf = pounds per cubic foot; psf = pounds per square foot * For a 2-by-4 wall cavity		

4.4.1 Preparing for Retrofit Wall Insulation

Inspect and repair walls thoroughly to avoid damaging the walls, blowing insulation into unwanted areas, or causing a hazard.

Inspecting and Repairing Walls

Before starting to blow insulation into walls, take the following preparatory steps.

- ✓ Calculate how many bags of insulation are needed to achieve the R-value specified on the bag's label. *See "Calculating Wall Insulation" on page 257.*

- ✓ Inspect walls for evidence of moisture damage. If condition of the siding, sheathing, or interior wall finish indicates an existing moisture problem, no sidewall insulation should be installed until the moisture problem has been identified and corrected.

- ✓ Seal gaps in external window trim and other areas that may admit rain water into the wall.

- ✓ Inspect indoor surfaces of exterior walls to assure that they are strong enough to withstand the force of insulation blowing. Reinforce siding and sheathing as necessary.

- ✓ Inspect for interior openings from which insulation may escape such as balloon framing openings in the attic or crawl space, pocket doors, un-backed cabinets, interior soffits, openings around pipes under sinks, and closets. Seal openings as necessary to prevent insulation from escaping.

- ✓ Confirm that exterior wall cavities aren't used as return or supply ducts. Avoid these cavities, or re-route the ducting.

- ✓ Ensure that electrical circuits contained within walls aren't overloaded. Maximum ampacity for 14-gauge copper wire is 15 amps and for 12-gauge copper wire is 20 amps. Install S-type fuses where appropriate to prevent circuit overloading. Don't insulate cavities containing knob-and-tube insulation. *See "Electrical Safety" on page 234.*

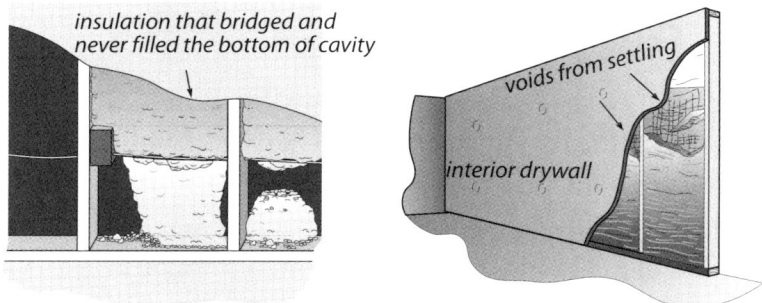

insulation that bridged and never filled the bottom of cavity

voids from settling

interior drywall

Problems with low density insulation: Blowing insulation through one or two small holes usually creates voids inside the wall cavity. This is because insulation won't reliably blow at an adequate density more than about one foot from the nozzle. Use tube-filling methods whenever possible, using a 1.5-inch hose inserted through a 2-inch or larger hole.

Drilling Exterior Sheathing: Insulation Retrofit

Avoid drilling through siding. Where possible, carefully remove siding and drill through sheathing. This avoids the potential lead-paint hazard of drilling the siding. It also makes it easier to insert flexible fill tubes since the holes pass through one less layer of material.

If the siding cannot be removed, consider drilling the walls from inside the home. Obtain the owner's permission before doing so, and practice lead-safe weatherization procedures. *See page 231.*

✓ Asbestos shingles may be carefully removed by pulling the nails holding them to the sheathing or else nipping off the nailheads. Dampening the asbestos tiles keeps dust down. Refer to your company standards for proper protective equipment when working with asbestos materials.

✓ Metal or vinyl siding may be removed using a zip tool.

✓ Homes with brick veneer or blind-nailed asbestos siding may be insulated from the inside. Holes drilled for insulation must be returned to an appearance as close to original as possible, or so they are satisfactory to the customer.

asbestos shingle

sheathing

end-cutting nippers

nails

Removing asbestos shingles: End-cutting nippers are used to pull the two face nails out of each shingle. Holes are then drilled in the sheathing for tube filling.

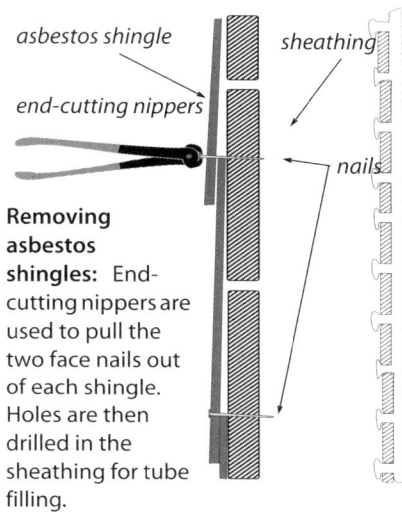

Removing metal siding: A zip tool separates joints in metal siding.

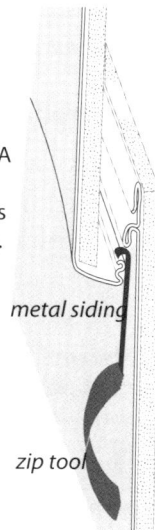

metal siding

zip tool

4.4.2 Retrofit Closed-Cavity Wall Insulation

This section describes three ways of installing wall insulation.

✓ Blowing walls with fibrous insulation using a fill tube from indoors or outdoors.

✓ Blowing walls with fibrous insulation from indoors or outdoors using a directional nozzle.

✓ Installing batts in an open wall cavity.

✓ Spraying wet-spray fiberglass or cellulose into an open wall cavity.

✓ Spray open-cell or closed-cell foam into an open wall cavity.

Installing Retrofit Fibrous Wall Insulation

Two methods for installing sidewall insulation are commonly used: tube-fill method (one large hole) or the multi-hole method, using a directional nozzle. The tube-fill method is preferred because it requires only one hole, and it ensures that wall achieves an adequate coverage and density of blown insulation.

Blowing Walls with a Fill-Tube

Dense-pack wall insulation is best installed using a blower equipped with separate controls for air and material feed. Mark the fill tube in one-foot intervals to help the installer verify when the tube has reached the top of the wall cavity.

Tube-filling walls: This method can be accomplished from inside or outside the home. It is the preferred wall insulation method because it is a reliable way to achieve a uniform coverage and density.

To prevent settling, cellulose insulation must be blown to at least 3.5 pounds per cubic foot (pcf) density. Fiberglass dense-pack must be 2.2 pcf and the fiberglass material must be designed for dense-pack installation. Blowing fibrous insulation this densely requires using a fill tube.

Insulate walls by this procedure.

1. Drill 2-to-3-inch diameter holes to access stud cavity.

2. Probe all wall cavities through holes, as you fill them with the fill tube, to identify fire blocking, diagonal bracing, and other

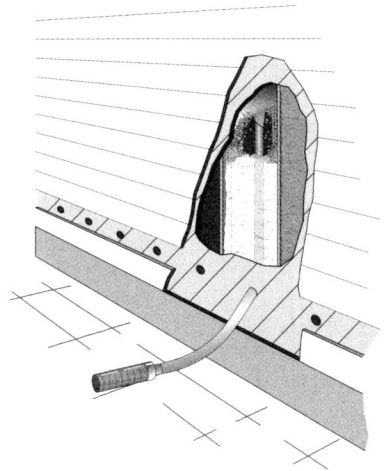

obstacles. After probing and filling, drill whatever additional holes are necessary to ensure complete coverage.

3. Start with several full-height, unobstructed wall cavities so you can measure the insulation density and calibrate the blower. An 8-foot cavity (2-by-4 on 16-inch centers) should consume a minimum of 10 pounds of cellulose.

4. Insert the hose all the way to the top of the cavity. Start the machine, and back the hose out slowly as the cavity fills.

5. Then fill the bottom of the cavity in the same way.

6. Learn to use the blower control to achieve a dense pack near the hole while limiting spillage.

7. Seal and plug the holes, repair the weather barrier, and replace the siding.

Insulation hoses, fittings, and the fill tube: Smooth, gradual transitions are important to the free flow of insulation.

When insulating balloon-framed walls, try to blow an insulation plug into each floor cavity to insulate the perimeter between the two floors. This also seals the floor cavity so it does not become a conduit for air migration. If the process is requiring too much insulation, try placing a plastic bag over the end of the fill tube and blowing the insulation into the plastic bag. The bag will limit the amount of insulation it takes to plug this area.

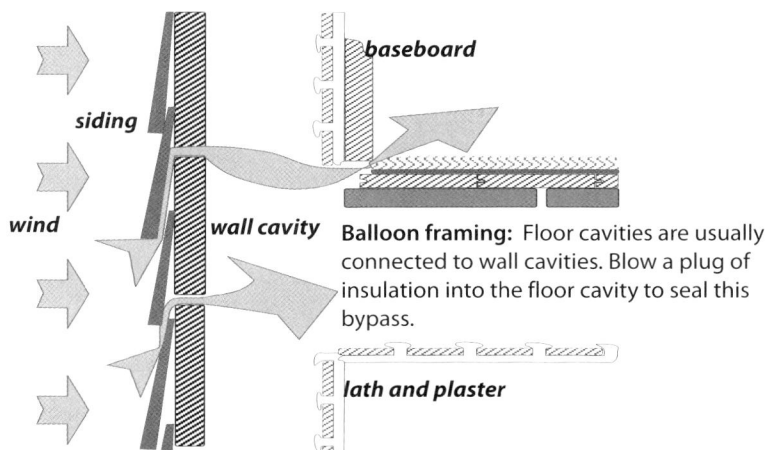

siding

wind

wall cavity

baseboard

Balloon framing: Floor cavities are usually connected to wall cavities. Blow a plug of insulation into the floor cavity to seal this bypass.

lath and plaster

Multi-Hole Blowing Method for Fibrous Insulation

The multi-hole method is often used when the insulator doesn't want to remove siding. The multi-hole method is the least preferable wall-insulation method because it often results in voids and sub-standard density. It is sometimes used effectively in conjunction with tube-filling to fill small cavities around doors or windows.

If you do employ this method, use a powerful blowing machine — a model designed for professional insulators.

✓ Drill holes into each stud cavity large enough to admit a directional nozzle. The holes should be spaced so that the insulation is blown no more than 12 inches upward or 24 inches downward.

Directional nozzle: It's difficult to achieve the correct density with the multi-hole method because the density decreases as the distance from the nozzle increases.

✓ Probe wall cavities to determine location of obstacles and nature of cavities around window and door areas.

✓ All wall cavities around windows and doors should be filled with insulation.

✓ Seal all holes with wood or plastic plugs.

Injecting Liquid Foam

Injecting liquid foam is more expensive than blowing fibrous insulation but offers better performance when existing walls are partially filled by batts. The batts are 1 to 2 inches thick and are usually flush with the interior drywall or plaster. The foam is best injected from outdoors to fill the cavity and compress the batt slightly. If injected from indoors, the foam may just stretch the batt facing and fail to create a full wall cavity.

Open-cell polyurethane foam, formulated to expand less than the sprayed variety, is the leading wall-retrofit foam. It must be installed through holes (<1 inch) spaced about two feet apart using a simple nozzle that barely enters the cavity. Technicians may use drinking straws or other indicators to judge the level that the foam has filled during installation. Technicians don't normally use fill tubes to inject open-cell foam because the tube would be too difficult to clean.

Air Krete or Tripolymer are both injected through a fill tube. Both these foams are water soluble allowing technicians to clean the fill tube after application. The fill tube allows technicians to remove a piece of siding and to fill a cavity from one hole (>2.5 inches).

4.4.3 Open-Cavity Wall Insulation

Fiberglass batts are the most common open-cavity wall insulation. They achieve their rated R-value only when installed carefully. If there are gaps between the cavity and batt at the top and bottom, the R-value can be reduced by as much as 30 percent. The batt should fill the entire cavity without spaces in corners or edges.

- ✓ Use unfaced friction-fit batt insulation where possible. Fluff the batts during installation to fill entire wall cavity.

- ✓ Choose medium- or high-density batts (R-13 or R-15 rather than R-11, and R-21 rather than R-19).

- ✓ Seal all significant cracks and gaps in the wall structure before or after the insulation is installed.

- ✓ Staple faced insulation to outside face of studs on the warm side of the cavity. Don't staple the facing to the side of the studs, even though drywallers may prefer that method, since it leaves an air space that encourages convection currents.

R-15 fiberglass batts: Install R-15 batts in open walls and attach them to the face of the stud as shown here. Or install unfaced batts. Either way, the batts should be accurately cut and carefully installed.

- ✓ Cut batt insulation to the exact length of the cavity. A too-short batt creates air spaces above and beneath the batt, allowing convection. A too-long batt will bunch up, creating air pockets.

- ✓ Split batt around wiring, rather than letting the wiring bunch the batt to one side of the cavity.

- ✓ Insulate behind and around obstacles with scrap pieces of batt before installing batt.

- ✓ Fiberglass insulation exposed to the interior living space must be covered with minimum half-inch drywall or other material that has an ASTM flame spread rating of 25 or less.

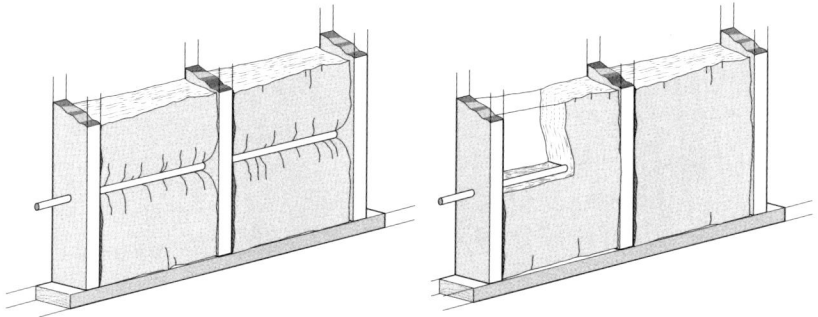

Fiberglass batts, compressed by a cable: This reduces the wall's R-value by creating a void between the insulation and interior wallboard.

Batt, split around a cable: The batt attains its rated R-value.

Sprayed Open-Cavity Wall Insulation

Both fibrous and foam insulation can be sprayed into open wall cavities. Varieties include the following.

- ✓ Fiberglass or cellulose mixed with water and glue at a special nozzle sprayed into the open wall cavity with the excess shaved off (fibrous damp-spray insulation)

- ✓ Open-cell or closed-cell polyurethane foam sprayed into open wall cavity and either held short of filling the whole cavity or with the excess foam shaved off after it cures

Spraying open-cell foam: The sprayed open-cell foam is left short of filling the cavity or else shaved off.

Blowing Open Wall Cavities behind Netting or Fabric

Blowing dry fibrous insulation behind netting or fabric is a common way of insulating open walls before drywall application, especially with cellulose. However, the insulation isn't usually blown to an sufficient density to resist settling. This method should be avoided unless the technician verifies at least 3.5 pcf for cellulose or 1.6 pcf for fiberglass — densities that can be difficult to achieve when blowing behind netting or fabric.

4.4.4 Insulating Deep Wall Cavities

Insulating deep wall cavities such as 2-by-8 walls and double 2-by-4 walls can be a challenge. Several common deep-wall insulation solutions present a high risk of large voids in the insulation. The following is a list of methods and the problems they may have with voids.

- ✓ Open-cell foam sprayed into the open wall — The foam has a potential for large bubbles and voids when blown in thicknesses beyond 4 inches.

- ✓ Blowing fibrous insulation behind netting or fabric in an open wall — Installer is unable to blow the insulation at a density sufficient to prevent settling.

- ✓ Fibrous damp-spray insulation installed in open wall — Walls become progressively more difficult to fill as the cavity gets deeper than a 2-by-6 wall. Slumping and a general failure of the wet insulation mass to fill the cavity and support itself in vertical cavity are a problem in deep walls.

Damp-spray cellulose in a double wall: Installers failed to fill this deep cavity due to slumping and shearing-off of the insulation. Drywall was applied to this assembly without further treatment

Blowing fiberglass through drywall: Installers blows fiberglass to 2.2 psf. Drywall installers leave a three-inch space and patch later with 3/8-inch drywall to leave a trough for drywall mud.

The following methods may avoid the problems described on the previous page.

- ✓ Designing a double wall to be insulated with high-density fiberglass or rock wool batts, which have less risk of slumping than less-dense batts. For example, use a 3.5 inch space between the walls and install batts horizontally in that space and vertically in the stud spaces of the two walls.

✓ Installing the drywall and then blow the cavity with fiberglass at 2.2 pounds per cubic foot density.

4.4.5 Insulated Wall Sheathing

Insulated sheathing is an excellent retrofit, especially when the siding and windows are being replaced. Insulated wall sheathing covers the wall surface with insulation, reducing thermal bridging through solid framing members.

Foam sheathing with battens: One-by-four battens are applied to 4 inches of foam board on the exterior provide a fastening strip for siding and trim.

Insulating wall sheathing is usually foam board, such as polystyrene or polyisocyanurate. Always fill the wall cavity before installing insulated sheathing.

Fastening Insulating Sheathing

Fastening the insulating sheathing requires one of the following to secure the insulation to the wood sheathing or masonry under it.

- ✓ A batten board
- ✓ An imbedded strip
- ✓ A broad staple
- ✓ A long screw with a large washer
- ✓ A special adhesive (masonry)

Foam with embedded strips: Strips of plywood or OSB are spaced on 16 or 24 inch centers at the factory. More strips or wide corner pieces can be added on the job with special foam cutting and grooving tools. Technicians install lightweight siding like steel or vinyl lap siding that drains well to protect the foam and strips.

Use appropriate fasteners for wood or masonry materials. Wood battens or embedded strips allow attachment of a variety of siding materials. The embedded strips work best with steel, aluminum, or vinyl siding, which are lightweight and drain water through weep holes in every unit of siding.

Exterior Insulation and Finish Systems (EIFS)

EIFS describes a number of proprietary systems for insulating and finishing building exteriors. Latex stucco can be troweled onto the surface of foam, fiberglass, or rock wool

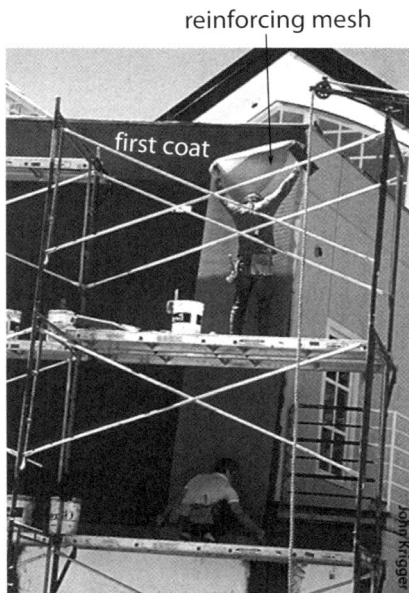

reinforcing mesh

first coat

John Krigger

EIFS system: Workers apply mesh reinforcing to a base coat of latex stucco material.

insulation to create EIFS. The exterior surface is a special stucco formulation that is usually less than $1/4$-inch thick. EIFS requires expert application and works best on wood-frame homes in dry climates and on masonry structures. EIFS may not be used to protect exterior foam insulation underground.

4.4.6 Wall Insulation within a Retrofit Frame

Retrofitters, seeking superior energy performance, sometimes build a wood-frame wall attached to the interior or exterior of the existing wall. Common insulation choices include all the wall choices discussed previously. Vapor barriers and air barriers must be incorporated into the new wall assembly as appropriate for the climate and existing wall characteristics. The exterior side of a retrofitted insulated frame should have sheathing and a water resistive barrier like house wrap.

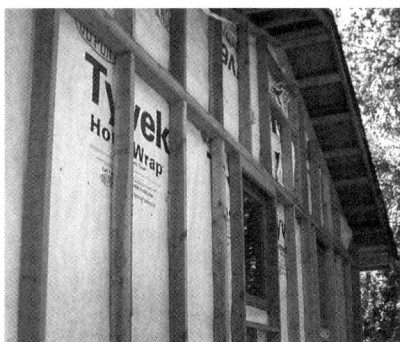

Frame wall for insulation: A frame wall is fastened with brackets that allow the wall to stand off the existing exterior wall by a half inch so foam can go behind the studs and plates to reduce thermal bridging through the assembly.

Open-cell or closed-cell polyurethane foam: The foam was sprayed into the cavities and the excess shaved off. New sheathing, house wrap, siding, and windows follow later.

4.4.7 Insulating Concrete Block Walls

Insulation can be installed in the cores of concrete blocks or to the interior or exterior surfaces. Insulate the cores with injectable foam or polystyrene beads. Foam products suitable for concrete block include the following.

✓ Tripolymer injectable foam. *See also "Tripolymer Injectable Foam" on page 100.*

✓ Air Krete injectable foam. *See also "Air Krete Injectable Foam" on page 100.*

✓ Open-cell polyurethane injectable foam. *See also "Open-Cell Polyurethane Spray Foam and Injectable Foam" on page 99.*

✓ Polystyrene beads.

Foam products that are suitable for surface application include the following.

✓ Polystyrene (EPS) foam attached to the block by special adhesive and covered with acrylic stucco as shown in

"Exterior Insulation and Finish Systems (EIFS)" on page 137.

✓ EPS or XPS foam equipped with strips or bonded to sheet goods to promote attachment and fire-safe finishes. *See also "Fastening Insulating Sheathing" on page 136.*

4.4.8 Insulating Brick Walls

There are three common wall types that incorporate brick as unreinforced masonry.

1. Traditional brick walls with header bricks that hold two layers of stretcher bricks together.

2. Various types of hollow brick walls with usually one layer of brick on either side of an air space.

3. Wood-frame brick veneer walls with a single layer of brick veneer that is attached to a typical wood frame wall.

Hollow brick wall: Two separate single brick walls are held together by wood lath embedded in the mortar joint.

All three of these assemblies may have structural problems depending on the age and details of construction. Consulting a structural engineer before making any modifications to these walls is essential. Mortar in traditional brick walls can turn to dust over the decades, old hollow brick walls can be frighteningly fragile, and small tremors can detach and tumble 100-year-old brick veneer. Be cautious with any type of retrofit involving these wall assemblies.

4.5 FLOOR AND FOUNDATION INSULATION

Floor and foundation insulation and air sealing complete the thermal boundary at the base of the building. This consideration is simply missing in many homes. In heating climates, floor and/or foundation insulation is a key part of improving the thermal performance of homes.

In homes with heated and occupied basements, the best choice is to insulate and air seal the basement walls, and so include the basement within the thermal boundary. The choice is less straightforward in homes with unused basements or crawl spaces where a choice must be made between insulating the floor or the foundation walls. This decision should be made according to cost-effectiveness and other factors as discussed in *"Decisions about Basement and Crawl Spaces" on page 55.*

To establish an effective thermal boundary, the insulation and air barrier should be adjacent to each other. Establishing an effective air barrier – comparable to the air barriers in the above-grade walls and ceiling – may be difficult. Furthermore, foundation or floor insulation may or may not be cost-effective or practical, considering the home's weatherization budget and potential moisture problems.

Most building experts prefer to insulate and air seal the foundation walls and not the floor because this strategy encloses the furnace, ducts, pipes and other features within an insulated and air sealed space. This involves plugging crawl-space vents if appropriate.

Floor insulation is generally preferred where there are crawl-space moisture problems or where rubble masonry makes insulating and air sealing the foundation wall difficult.

4.5.1 Preparing for Floor or Foundation Insulation

Floor and foundation insulation can increase the likelihood of moisture problems. Installers and home owners should take all necessary steps to prevent moisture problems from ground moisture sources.

Moisture Source-Reduction

Observe the following specifications for avoiding the deteriorating effects of crawl-space and basement moisture on insulation and other building materials.

✓ Solve all drainage problems, ground-water problems, wood-deterioration, and structural problems before installing floor or foundation insulation.

✓ Slope the ground outside the home away from the foundation.

✓ Install gutters and downspouts in wet locations and direct roof water away from the home.

✓ Install a ground-moisture barrier in all dirt-floored crawl spaces.

✓ Confirm that all combustion vents (chimneys), clothes-dryer vents, and exhaust fan vents are vented to outdoors.

Sump pump: These pump water out of a sump or basin where water collects in a basement or crawl space.

✓ Suggest a sump pump for crawl spaces or basements with a history of flooding. The sump pump should be located in an area where it will collect water from the entire below-grade area and pump it to a drain or swale outdoors away from the foundation.

Ground Moisture Barriers

Air, water vapor, liquid water, and pollutants move through soil and into crawl spaces and dirt-floor basements. Even soil that seems dry and airtight at its surface can convey a lot of water,

water vapor, and soil gases into a home. For these reasons, all crawl spaces should have an airtight ground moisture barrier.

Cover the ground in the crawl space with an airtight moisture barrier to prevent the movement of moisture and soil gases from the ground into the crawl space.

- ✓ Cover the ground completely with a ground moisture barrier such as 6-to-10-mil polyethylene. Reinforced or cross-linked polyethylene is more durable than un-reinforced polyethylene.

- ✓ Run the moisture barrier up the foundation wall several inches and fasten it with polyurethane adhesive or acoustical sealant to a clean and flat masonry surface.

- ✓ Seal the edges and seams with urethane, acoustical sealant, butyl caulking, or construction tape to create an airtight seal between the crawl space and the ground.

- ✓ To prevent conveyance or trapping of moisture against wood surfaces, ground moisture barriers must not contact wood structural members, such as posts, mud sills, or floor joists.

Naturally Ventilated Crawl Spaces

When insulating the floor, the crawl space is usually ventilated naturally through passive vents. A ground moisture barrier is required to protect the floor insulation and other building materials throughout the home from moisture.

- ✓ A crawl-space with a ground-moisture barrier may have vent openings equal to 1 square foot of vent area to 300 square feet of crawl-space floor area. A minimum of two vents should be installed on opposite corners of the crawl space.

- ✓ In a dry crawl space with a ground-moisture barrier, ventilation openings may be minimized to one square foot of net free ventilation area for every 1500 square feet of crawl-space floor area, according to the 2009 IRC.

Power-Ventilated or Conditioned Crawl Spaces

Most code jurisdictions allow the elimination of passive crawl-space vents when the foundation walls will be insulated. Consider these common specifications.

- ✓ If local codes allow and you have effectively reduced moisture sources, installed a ground-moisture barrier, then seal the passive foundation vents completely.

- ✓ The 2009 International Residential Code (IRC) requires insulation installed from the subfloor to the ground in the crawl space and then run the insulation 24 inches horizontally to lengthen the horizontal heat transmission path from the crawl space to outdoors.

- ✓ The 2009 International Residential Code (IRC) requires 1 CFM per 50 square feet of crawl space floor area in powered exhaust ventilation or the same amount of conditioned supply air from a forced-air system. In either of these options, the IRC requires openings from the crawl space into the home.

The conditioned crawl space, although allowed by the IRC, is a questionable moisture-and-energy solution for crawl spaces. While heating the ground in winter is merely a waste of energy, refrigerating the ground in summer is both wasteful and a potential cause of moisture problems. Reducing the temperature of floor joists and other building materials may raise their moisture content and lead to surface condensation.

Rim Insulation and Air Sealing

The joist spaces at the perimeter of the floor are a major weak point in the air barrier and insulation. Insulating and air sealing both the rim joist and longitudinal box joist are appropriate either as individual procedures or as part of floor or foundation insulation.

air seal stud cavities in balloon-framed homes as a part of insulating the rim joist. air seal other penetrations through the rim before insulating. Two-part spray foam is the most versatile air sealing and insulation system for the rim joist because spray foam air seals and insulates in one step. Polystyrene or polyurethane rigid board insulation are also good for insulating and air sealing the rim joist area. When the rim joist runs parallel to the foundation wall, the cavity may be air sealed and insulated with methods similar to those as shown here.

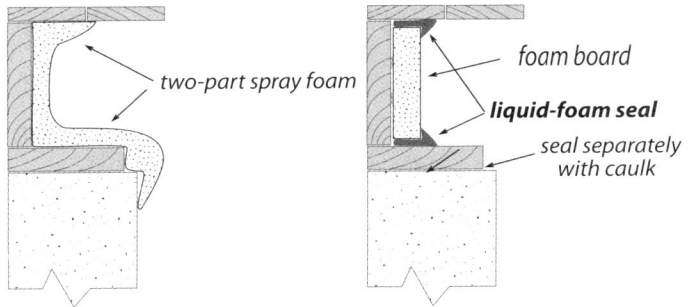

Foam-insulated rim joists: Installing foam insulation is the best way to insulate and air seal the rim joist.

Don't use fiberglass batts to insulate rim joist areas because air can circulate around the fiberglass, causing condensation and encouraging mold on the cold rim joist. If foam-board is used to insulate the rim, liquid foam sealant should be used to seal around the edges.

4.5.2 Installing Floor Insulation

The best way to insulate a floor cavity is to completely fill each joist cavity with fiberglass insulation. Blowing fiberglass insulation is the easiest way to achieve complete coverage because the blown fiberglass is better able to surround obstructions and penetrations than fiberglass batts. Before installing fiberglass floor insulation, make the following preparations.

✓ Establish an effective air barrier at the floor, prior to insulating the floor, to prevent air from passing through or around floor insulation.

✓ Seal and insulate ducts remaining in the crawl space or unoccupied basement

✓ Insulate water lines if they protrude below the insulation.

Blowing Floor Insulation

Many installers prefer to blow fiberglass insulation into the floor, rather than installing batts, to insulate more easily and effectively around obstacles. Cellulose is not recommended for blowing floors because of concerns about settling and moisture. Observe these specifications for blown-in floor insulation.

✓ The entire under-floor surface must be covered by a vapor permeable air barrier: insulation dust-free mesh, Tyvek®, perforated $1/4$-inch foam board, or equivalent material.

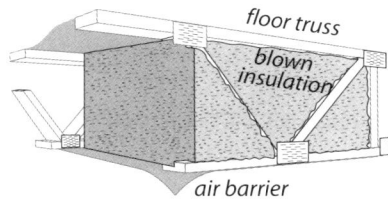

Blown floor insulation: This method works particularly well with floor trusses.

✓ A flexible or semi-flexible air barrier or containment barrier must be supported by twine or wood strips unless specifically designed to support floor insulation without reinforcement.

✓ Twine must be fastened with staples at least $5/8$ inches in length and made of 18-gauge corrosion-resistant metal.

✓ Install rock wool or fiberglass blowing wool through V-shaped holes in the air barrier.

✓ Use a fill tube for installing the blown insulation. Insulation must travel no more than 12 inches from the end of the fill tube to its final destination.

✓ All penetrations in the air barrier must be sealed with a tape, approved for sealing seams in the air-barrier materia

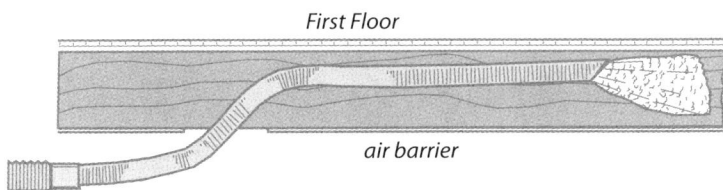

First Floor

air barrier

Blowing floor cavity: Uninsulated floor cavities can be blown with fiberglass or rock wool insulation, using a fill tube.

All penetrations in the air barrier must be sealed with a tape, approved for sealing seams in the air-barrier material

Installing Batt Floor Insulation

Observe the following specifications for insulating under floors.

1. Seal all significant air leaks through the floor before insulating the floor, using strong airtight materials.

2. Use fiberglass batts or dense-packed blown fiberglass insulation. Avoid blown cellulose because of its weight, moisture absorption, and tendency to settle.

3. Batts should be in continuous contact with the subfloor.

4. Batts must either fill the complete depth of each cavity or have a thermal resistance of at least R-30.

5. Batts must be neatly installed, fitting tightly together at joints, fitting closely around obstructions, and filling all the space within the floor cavity.

6. Crawlspace access doors, adjacent to a conditioned space, must be insulated to at least R-21 for horizontal openings and to at least R-15 for vertical openings.

7. Crawlspace access doors must be effectively weather-stripped.

Fasteners for floor insulation must resist gravity, the weight of insulation, and moisture condensation.

1. Batt insulation, installed in floors, must be supported by twine, wire, or wood lath and must be in contact with the floor.

   a. Twine must be made of polypropylene or polyester and have a breaking strength of at least 150 pounds.

   b. Wire must be copper or stainless steel with a minimum diameter of 0.04 inches or size 18 AWG.

   c. Twine or wire must be installed in a zig-zag pattern. The staples must be no more than 12 inches apart for joists on 24-inch centers and 18 inches apart for joists on 16-inch centers.

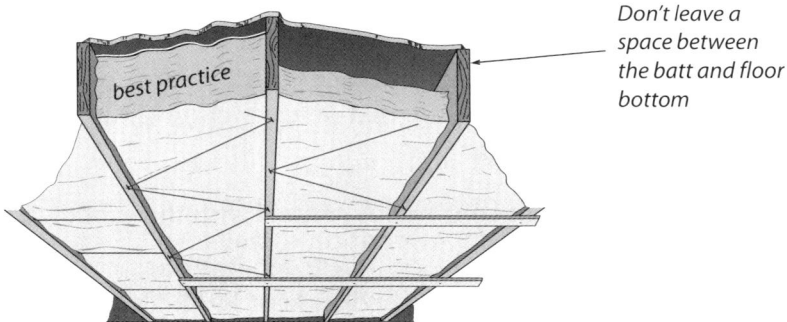

best practice

Don't leave a space between the batt and floor bottom

Floor insulating with batts: Use unfaced fiberglass batts, installed flush to the floor bottom, to insulate floors. The batt should fill the whole cavity if it is supported by lath or plastic twine underneath. For batts that don't fill the whole cavity, use wire insulation supports.

4.5.3 Crawl-Space Foundation Insulation

Crawl-space foundation is a retrofit worthy of consideration in heating climates. This retrofit is only worthwhile if you can seal the existing foundation vents. *See "Naturally Ventilated Crawl Spaces" on page 142.*

Foundation insulation is usually installed on the inside of the foundation walls during retrofits. Exterior insulation is common in new construction because it is more effective than interior insulation at keeping the foundation warm and dry. Don't insulate foundation walls on the interior unless the crawl space has a ground-moisture barrier and you confirm that no drainage problems exist.

Fiberglass blankets can be a good insulation choice but the facing, if present, should be vapor permeable and not a vapor barrier. A vapor barrier like vinyl or foil can trap moisture in the insulation for extended time periods.

Extruded or expanded polystyrene insulations are the most appropriate insulation products for the interior of flat concrete or concrete-block walls. Foam sheets are good air barriers with excellent moisture resistance. For rubble masonry walls, use two-part spray foam. Spray foam is preferred whenever there is much moisture present. Consider the following issues when insulating foundation walls.

✓ If an open-combustion appliance is located in crawl space, take precautions to assure that outdoor combustion air is available to the appliance.

✓ When insulating crawl-space walls, check with the local building official about ventilation options.

✓ Outside access hatches should be securely attached to foundation wall. If the foundation walls are insulated, any crawl-space access hatch from outdoors should be insulated.

✓ In regions affected by termites, carpenter ants, and similar insects, consult with experts to ensure that the insulation, air sealing materials, and moisture barrier don't provide a conduit for insect infestation into wood materials.

Wet-spray fiberglass: Wet-spray fiberglass is acceptable for very dry crawl spaces. It can also be used as an ignition barrier to cover plastic-foam insulation.

Insulating the foundation walls instead of the floor usually leaves a large area of the homes thermal boundary — the ground in the crawl space — uninsulated. This condition results from the erroneous but widespread assumption that winter heat loss through the ground is negligible.

Foundation-Insulation Materials Compared

Crawl-space foundation insulation is rather complex from the perspective of materials. Foam insulation has special fire considerations. Faced fiberglass can trap moisture. Sprayed fiberglass can be damaged by moisture and mechanical abrasion. Consider the following notes about crawl-space foundation insulation materials.

✓ Attach insulation firmly to the entire inside wall surface with appropriate fasteners and/or adhesive. Install insulation with no significant voids or edge gaps.

✓ Plastic foam insulation is combustible and creates a smoke hazard when it burns. Some code jurisdictions allow it to be installed uncovered in crawl spaces and some jurisdictions require it to be covered by a ignition barrier.

✓ Sprayed closed-cell polyurethane foam is available in at least two varieties according to flame spread and smoke developed fire ratings. Use the more fire-safe variety if available.

subfloor

rim joist

two-part foam

sheet foam or other insulation

ground-moisture barrier

foundation wall

footing

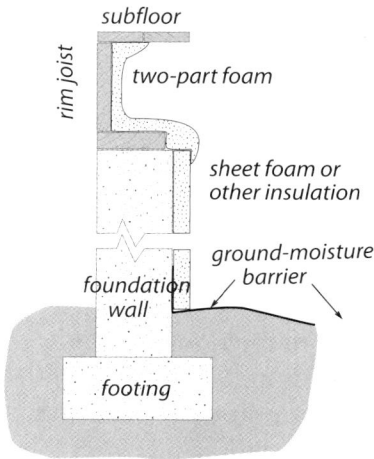

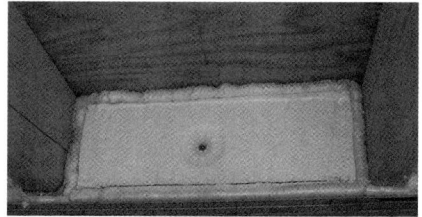

Foam-insulated rim joists: Two-part foam insulates and seals the rim joist and sheet foam insulates the foundation walls (left).
(Above) Four-inch polystyrene foam is cut an inch short in both dimensions and attached to a rim joist with a long screw and large plastic washer. One part foam fills the gap around the foam rectangle.

✓ When an ignition barrier is required, foam insulation must be protected by $1^1/_2$ inch thick of fiberglass insulation or a proprietary sprayed ignition barrier.

✓ Sprayed fiber insulation is acceptable as an insulating material, but unlike spray foam, isn't an air barrier. However, the spray fiber does serve as an ignition barrier over the foam. Seal the rim joist area from the subfloor to the concrete below the sill with spray foam before spraying the fiber.

✓ Choose fiberglass rather than cellulose damp-spray because of its lower moisture absorption.

✓ FSK (foil scrim kraft) or vinyl-faced fiberglass insulation is often used to insulate foundation walls on the interior surface. Both of these facings are vapor barriers and can trap moisture inside the insulation.

✓ Cover basement foundation insulation with a material that has an ASTM flame spread rating of 25 or less, such as half-inch drywall. FSK fiberglass usually requires no ignition barrier.

Labels in left illustration: foam sealant, foam rim joist insulation, drywall, strip, beadboard

Labels in right illustration: floor joist, stud wall, spray foam, custom flashing

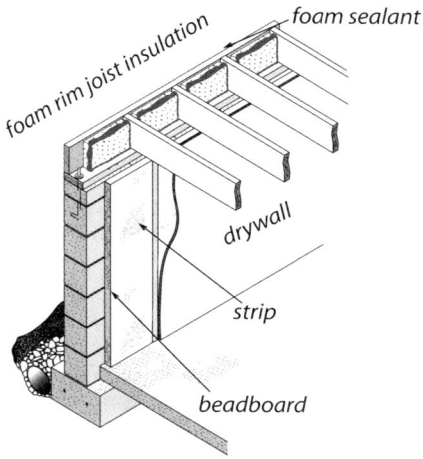

2-inch foam board with plywood strips: The built-in strips facilitate fastening the foam to the wall and fastening drywall to the foam.

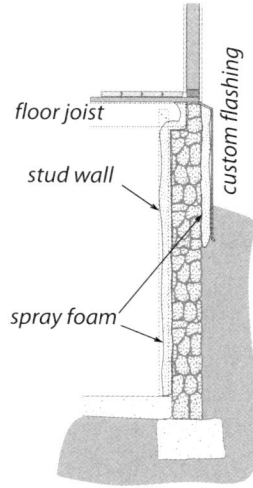

2-part foam sprayed on rubble masonry: Rubble masonry walls can be insulated on the interior or exterior with sprayed plastic foam. On the interior, the foam must be covered by drywall.

4.5.4 Basement Insulation

Basement wall insulation is a problematical retrofit for heating climates. It is seldom executed properly because of poor understanding about moisture transport mechanisms.

Frame-Wall Insulation

The most common (although not the best) way to insulate basement walls, or any masonry wall, is to build a framed wall against the masonry wall and fill it with fiberglass batts. The frame is then covered with drywall.

Frame-wall method: A common method, but it's seldom very effective and often leads to moisture damage, especially if the wall has a vapor barrier. This method can be acceptable with meticulous air sealing to prevent basement air from circulating behind the wall. Careful exterior drainage control is also absolutely necessary.

Unfaced batts are probably the best choice of fiberglass insulation since they contain no vapor barrier to trap moisture. Moisture may travel in either direction: from outdoors in or indoors out.

With a framed wall, the installer often neglects to seal in areas where the wall is discontinuous, such as a mechanical room. Any areas, such as unfinished wall, open rim-joist area, or un-sheeted ceiling constitutes a very large thermal bypass. Avoid this problem by following these specifications.

- ✓ Insulate the rim joist and air seal it.
- ✓ Build the frame wall and stand it up.
- ✓ Wall off the entire basement. If a mechanical room or other area will not be insulated, install an airtight block at the wall's edge to prevent basement air from circulating behind the insulated wall.

✓ Don't install a vapor barrier on the interior face of the new basement wall. This will allow the wall to dry toward indoors or outdoors.

✓ Install drywall in an airtight manner on the walls and ceiling by applying sealant to the back or each sheet around its perimeter.

Stripped Foam Insulation

An excellent choice for insulating flat basement walls is polystyrene foam, installed in sheets at least 2 inches thick.

You can order either expanded polystyrene or extruded polystyrene equipped with grooves for fastening strips on 2-foot or 16-inch centers. This may be the easiest and most satisfactory way of insulating below-grade walls. Follow these specifications for installing 2-inch stripped foam.

Installing stripped foam: The foam is glued and screwed to the masonry wall. The drywall is glued to the foam and screwed to the strips. Electrical boxes are installed with the insulation.

✓ Apply walnut-sized globs of adhesive to the back of the sheet on one-foot centers. Use a foam-compatible adhesive and follow the instructions on the container.

✓ Install at least two concrete screws or two powder-driven nails in each strip, 24 inches from the bottom and top.

✓ Wherever an outlet is needed, install it between two sheets. Cut the rectangle out of one of the sheets. Install a two-inch plastic box backed by a piece of half-inch plywood, using construction adhesive and a concrete screw.

✓ Leave a half-inch gap at the bottom of the polystyrene sheets to run wire. Run the wire along the floor and up into the boxes.

✓ Seal the bottom gap and other gaps in the foam sheeting with one-part foam.

✓ Glue the drywall using the same adhesive and pattern. Screw the drywall to each wooden strip with one-inch drywall screws.

Exterior Foam Insulation

If installed at the exterior, as during new construction, use durable water-resistant insulation such as blue or pink extruded polystyrene or high-density (2 pcf) expanded polystyrene. For portions that are exposed above ground level and six inches below ground, you'll need to provide mechanical and moisture protection such as sheet metal or fiberglass panels. For areas more than 6 inches below grade, there are asphalt-based sealants for the foam that are applied with a paint roller.

4.6 Duct Insulation

Insulate supply ducts that run through unconditioned areas outside the thermal boundary such as crawl spaces, attics, and attached garages with a minimum of R-6 vinyl- or foil-faced duct insulation. Don't insulate ducts that run through conditioned areas unless they cause overheating in winter or condensation in summer. Follow the best practices listed below for installing insulation.

✓ Always perform necessary duct sealing before insulating ducts.

✓ Duct-insulation R-value must meet or exceed IECC 2009 standards: R-8 for vented attics, R-6 for other areas.

✓ Insulation should cover all exposed supply ducts, without significant areas of bare duct left uninsulated.

✓ Insulation's compressed thickness must be more than 75% of its uncompressed thickness. Don't compress duct insulation excessively at corner bends.

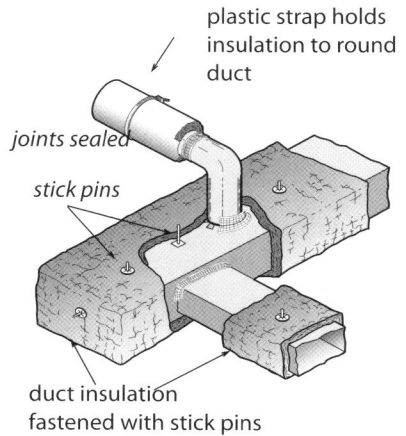

✓ Fasten insulation using mechanical means such as stick pins, twine, staples, or plastic straps.

plastic strap holds insulation to round duct

joints sealed

stick pins

duct insulation fastened with stick pins

Duct insulation: Supply ducts, located in unheated areas, should be insulated to a minimum of R-6.

Tape is essential for covering joints in the insulation to prevent air convection and the condensation that could result from convection. However, tape often fails as a supporting material if expected to resist the force of the insulation's compression or weight.

Caution: Burying ducts in attic insulation is a common practice in some regions and clearly reduces energy losses from ducts. However, burying ducts in attic insulation can cause condensation on ducts in humid climates.

CHAPTER 5: WINDOWS AND DOORS

This chapter presents specifications and procedures for improving the energy performance of windows and doors. Detailed specifications for window replacement provide guidance on this often-performed and potentially troublesome retrofit.

Windows and doors are a major concern to homeowners and energy specialists alike. Windows and doors were once thought to be a major Air Leakage problem. However, the widespread use of blower doors has shown that windows and doors don't tend to harbor large air leaks. But the combined heat losses and gains, by conduction, convection, radiation, and air leakage through windows are usually very significant.

Unfortunately, the square-foot cost to improve windows is high, so the payback from window improvements is usually not as attractive as many other building-shell retrofits. In older buildings, though, the windows and doors may be in such poor condition that repair or replacement is essential to a building's survival even if it's not an energy-saving priority.

All tasks relating to window and door improvement, repair, and replacement should be accomplished using lead-safe weatherization methods. *See "Lead-Safe Procedures" on page 231.*

5.1 WINDOW SHADING

Between 35% to 85% of the solar energy that strikes a home's windows passes through the glass and enters the living space. Solar heat may be welcome in cold weather, but solar heating through windows accounts for up to 40% of summer overheating in many homes. To reduce cooling costs, block solar heat before it enters the home.

Window shading increases comfort and reduces the cost of cooling, and is one of the most cost-effective weatherization measures in hot climates. Not all windows cause overheating, so you should prioritize your efforts towards windows where the most heat enters including the following.

- ✓ Windows that face east or west and are exposed to low-angle sun
- ✓ South side windows without adequate overhangs
- ✓ Windows with no shade from landscaping or roof over-hangs
- ✓ Large windows

5.1.1 Outdoor Window Shading Strategies

Solar screens, awnings, shutters, or rolling shades can provide good window shade. Exterior shading devices are inherently more effective than interior shading devices.

Solar Screens

Solar screens, made of mesh fabric that is stretched over an aluminum frame, are one of the most effective window shading options. They absorb or reflect a large portion of the solar energy that strikes them, while allowing a slightly diminished but acceptable view out of the window.

Solar screens are installed on the outside of the window, and work well on fixed, double-hung, or sliding windows. They aren't suitable for jalousie windows. For casement and awning windows, the sun screen should be mounted on the movable window sash rather than on the window frame. Sun screens are easily constructed in the shop or on the job.

spline

fabric

frame

Sun screen

Rolling the spline

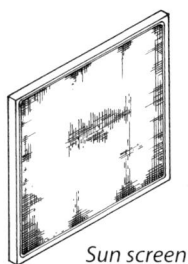

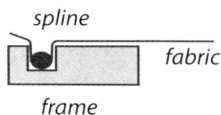

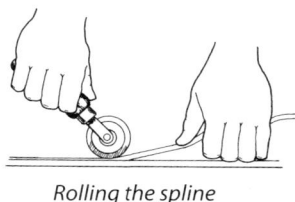

Sun screens: Installed on the window's exterior, sun screens absorb solar heat before it enters the home. This strategy is superior to interior window treatments, which reflect heat back after it has entered.

- ✓ Cut the frames to size using a metal cutting saw.
- ✓ Fasten the frames together through reinforced corner pieces.
- ✓ Cut the fabric a few inches large in length and width, and stretch it into the frame using continuous soft plastic splines that fit into the frame.
- ✓ Cut the excess fabric from around the edges of the channel.

Install solar screens on the exterior of the window frame, trim, or sash. Drill pilot holes for screws that pass through the aluminum frame, or use clips that are screwed to the window frame outside the sun screen. Use aluminum fasteners on aluminum frames to avoid corrosion.

Window Films

Metallized plastic window films, similar to those applied to automotive windows, can block 50 to 75 percent of the solar heat that comes through single-pane glass. A microscopic layer of metal on the film reflects solar radiation.

The material comes in rolls, and is permanently applied to the glass. Installed on the interior side of single-pane or double-pane glass, reflective window films repel solar heat (reduce solar heat gain), reduce glare, and reduce wintertime heat loss. Films also reduce UV light, which protects fabrics and carpets from fading, and they make the glass shatter resistant. Tinted films that merely color the glass are not very effective at blocking solar heat. Clear, spectrally selective films can block up to 70 percent of the solar energy while stopping only 30 percent of the visible light.

Installing reflective window film requires experience and/or training with planning, cleanliness, and care.

Awnings

Awnings are usually the most expensive window shading option. However, they are very effective at shading because they intercept the solar heat before it gets to the window. Awnings are popular in hot, sunny climates.

The amount of shade that an awning will cast over a window is determined mostly by the distance that it protrudes down from the top of the window. This distance is referred to as the "drop" of the awning. Awnings on the south side need a drop measuring 45 to 60 percent of the window height to block solar radiation coming from higher in the sky. Awnings on the east and west should have a drop of 60 to 75 percent to block solar radiation coming from lower in the sky.

Awnings with sides in addition to a top provide the most effective shade on south facing windows. Do-it-yourself awnings usually do not have sides but you can partially compensate for this by making the awning wider than the window. The greater the drop of the awning the more the view is reduced, so strike a balance between shade and view.

Metal awnings have the longest lifespan, though many homeowners feel that canvas and other fabric awnings are more attractive. Fabric is also more expensive and slightly less effective because it absorbs more solar heat. Fabric awnings need more maintenance, too.

It can be difficult to strike a perfect balance among the important factors of shade, view, appearance, and cost. Awnings for tall narrow windows may need to be considerably wider than the window for a more balanced appearance. Awnings installed at a 45-degree angle seem most attractive. If a maximum amount of view is important, use sun screens and window

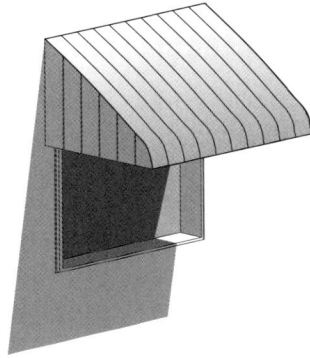

Aluminum awning

Retractable fabric awning

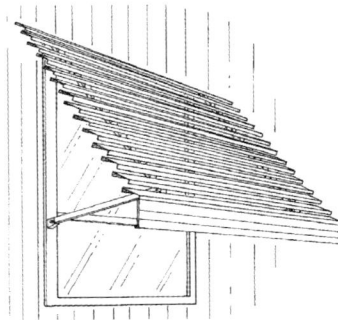

Slatted awning

films instead of awnings. Awnings with slats rather than a solid surface will allow some limited viewing through the top of the window.

Landscaping for Shade

Trees and bushes can provide shade for windows, walls, and roofs. Trees also cool the air around the home with shade and moisture evaporating from their leaves. Well-planned landscaping can reduce an un-shaded home's air conditioning costs by up to 50% while adding value to the home. Consider solar water heating and photovoltaic electricity before developing a landscaping plan.

The best plan for cool landscaping includes tall deciduous trees on the south side of the home to block high mid-day sun. Shorter trees or bushes on the east and west block morning and afternoon sun. Keep bushes far enough away from the foundation so that watering doesn't cause moisture problems.

Plant deciduous trees that lose their leaves in the autumn to admit winter sun. Choose types that are quick-growing and easy to care for in your region. Check with a local nurseryman to determine the best varieties of trees, when to plant, and the planting method.

Trees for shade: Landscaping is a good long-term investment for residences. Tall deciduous trees on the south block high summer sun while allowing lower winter sun to reach the home. Shorter trees or bushes provide protection from low-angle sun on the east and west.

5.1.2 Indoor Window Shading Strategies

Venetian Blinds

Venetian blinds can be installed on either the interior or exterior of your windows. They regulate both natural light and solar heat and also block glare with their movable slats. When designed for controlling light and solar heat, the slats should be bright white or polished aluminum.

When closed, interior venetian blinds can block up to 60 percent of the solar heat that would normally enter a single-pane window.

Roller Shades

Modern roller shades are made in a wide range of styles, including some that are very effective at reflecting and blocking solar heat.

Roller shades come in a wide variety of woven fabrics with different shading capacity. The exterior-facing side of the shade should be bright white or have a reflective metallized coating if they are installed for blocking solar heat.

New specialized shade fabrics are designed to preserve the view, prevent glare, and block solar heat. These innovative fabrics are colored on the indoor side for glare-free outdoor viewing, and white or shiny metallic facing outdoors to reflect solar heat.

If you have a substantial overhang to protect outdoor shades from the weather, you can install a large roller shade on the exterior to shade large windows or sliding-glass doors.

South windows admit solar heat during the middle of the day, coming from high in the sky. If the building has overhangs, the bottom of the south facing window glass needs shading more than the top. In this case, consider mounting the shades at the bottom of the window to provide shade at the window's bottom while allowing a view and natural light through the window's shaded top.

Operating Interior Window Treatments

Operating moveable window treatments correctly is essential to their effectiveness. Discuss these principles with your customers.

- ✓ Close window shades in the morning before the home begins to heat up. Close the windows at that time as well.
- ✓ Open shades in the evening to help cool the home. Open the windows if appropriate and safe.
- ✓ Open shades all day during winter to allow solar heat to enter the home.

5.2 STORM WINDOWS

Storm windows may be installed on the interior or exterior of existing primary windows. The storm windows may be operable or fixed and may be installed on either the window frame or window sash.

5.2.1 Exterior Aluminum Storm Windows

Storm windows are less expensive than window replacement. Storm windows can preserve old worn primary windows from more destructive weathering.

Aluminum storm windows are the best choice if they are well designed and installed properly

- ✓ Frames should have sturdy corners so they don't rack out-of-square during transport and installation.
- ✓ Sashes must fit tightly in their frames.
- ✓ The gasket sealing the glass should surround the glass's edge..
- ✓ The storm window should be sized correctly and fit tightly in the opening.

The installation of storm windows should follow these guidelines.

- ✓ Don't install new storm windows to replace existing storms if the existing storms are in good condition or can be repaired at a reasonable cost.

- ✓ Caulk storm windows around the frame at time of installation, except for weep holes at the bottom that must not be sealed. If weep holes are not manufactured into new storm window, they should be drilled.

- ✓ Don't allow storm windows to restrict emergency egress or ventilation through moveable windows. Choose windows that are openable from the inside, or install pin-on storm sashes that open along with the moveable primary window.

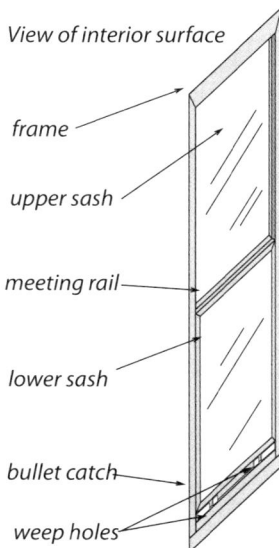

View of interior surface

frame

upper sash

meeting rail

lower sash

bullet catch

weep holes

Aluminum exterior storm windows: They protect the primary window and add about an R-1 to the window assembly.

5.2.2 Interior Storm Windows

Interior storm windows must have an effective perimeter air seal to prevent the warm, moist, indoor air from depositing condensation on the cool primary window during the winter.

Many interior storm windows have flexible or rigid plastic glazing that gives them a slightly higher R-value than glass interior storms.

It's important to look closely at the frames, sashes, and seals of the interior storm window assembly to evaluate its durability. Be skeptical about magnetic tape and other components that depend on adhesives as their sole means of attachment to the existing window frame.

Flexible storm with magnetic seal

Installing an inexpensive metal primary window at the interior side of the exterior primary window is a popular treatment for creating double glazing in hotels and apartment buildings. Mobile homes also use the strategy of having a secondary operable window inside the exterior primary window. *See "Mobile Home Storm Windows" on page 201.*

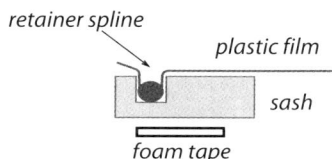

retainer spline

plastic film

sash

foam tape

Lightweight interior storm windows: Glass is the most durable glazing for interior storm windows. However a variety of flexible plastic glazings are also available for less cost

5.2.3 Storm Window Panels

Storm window panels combine a glazing material, usually glass, with a sash material such as aluminum or plastic. These fixed panels can be added to the interior or exterior of existing windows to improve their thermal resistance. If a storm panel is installed on the interior side of a window its frame would be sealed with foam tape or another method to prevent the intrusion of warm moist air into the space between the panel and existing window.

Storm window panels can be attached to the existing window frame or window sash. Existing windows that are operable must not be rendered inoperable by the installation of storm panels.

Storm panel: Storm panels add a pane of glass, preferably low-e, to an existing window for a very reasonable cost.

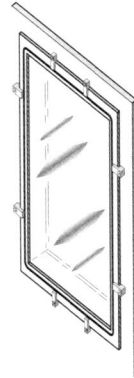

Low-e glass is a good choice for glazing storm window panels. However, the low-e surface isn't as easy to clean as a normal glass surface. Therefore the low-e surface should face another pane of glass in order to keep it clean

5.3 WINDOW REPAIR AND AIR LEAKAGE REDUCTION

With the exception of broken glass or missing window panes, windows are rarely the major source of air leakage in a home.

Window weatherstripping is typically not cost-effective but may be installed to solve a comfort problem. Avoid expensive or time-consuming window repair measures that are implemented to solve minor comfort complaints.

Re-glazing window sashes is time consuming, and is best accomplished as part of a comprehensive window rehabilitation project. Re-glazing wood windows may not be a durable repair without thorough scraping, priming, and painting. Repair measures may include the following measures.

✓ Replace missing or broken glass. Use glazing compound and glazier points when replacing glass in older windows. Glass cracks that are not noticeably separated can be ignored.

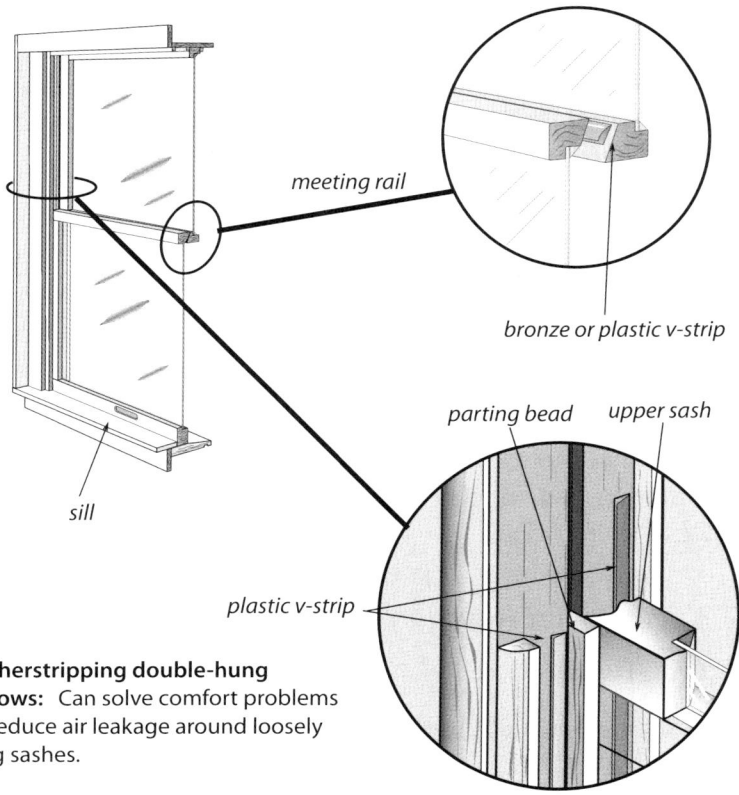

meeting rail

bronze or plastic v-strip

parting bead   upper sash

plastic v-strip

Weatherstripping double-hung windows: Can solve comfort problems and reduce air leakage around loosely fitting sashes.

sill

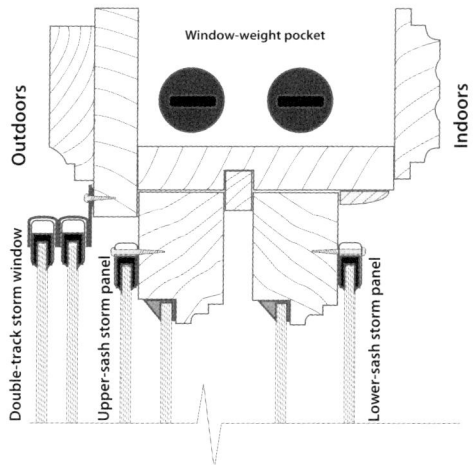

Outdoors

Window-weight pocket

Indoors

Double-track storm window

Upper-sash storm panel

Lower-sash storm panel

Optimized double-hung window: An exterior aluminum storm window plus storm window panels on the window sashes create triple glazing in this double-hung window.

- ✓ Caulk interior and exterior window frame to prevent air leakage, condensation, and rain leakage. Use sealants with rated adhesion and joint movement characteristics appropriate for both the window frame and the building materials surrounding the window.

- ✓ Replace missing or severely deteriorated window frame components. Extremely damaged wood should filled with a marine epoxy, primed, and painted.

- ✓ Adjust window stops if large gaps exist between stop and jamb. Ensure that the window operates smoothly following stop adjustment.

- ✓ Weatherstrip large gaps between the sash and the sill or stops. Weatherstrip the meeting rails if needed.

- ✓ Replace or repair missing or non-functional top and side sash locks, hinges, or other hardware if such action will significantly reduce air leakage.

Use lead-safe work practices when working on windows. *See page 231.*

Weatherstripping Double-Hung Windows

Wooden double-hung windows are fairly easy to weatherstrip. Keep in mind, window weatherstripping is mainly a comfort retrofit and a low weatherization priority.

Paint is the primary obstacle when weatherstripping double-hung windows. Often the upper sash has slipped down, and is locked in place by layers of paint, producing a leaking gap between the meeting rails of the upper and lower sashes.

- ✓ To make the meeting rails meet again, either break the paint seal and push the upper sash up, or cut the bottom of the lower sash off to bring it down.

- ✓ To lift the upper sash, cut the paint around its inside and outside perimeter. Use leverage or a small hydraulic jack to lift the sash. Jack only at the corners of the sash. Lifting in the middle will likely break the window.

- ✓ Block, screw, or nail the repositioned upper sash into place.

- ✓ To weatherstrip the window, you must remove the lower sash. Cut the paint where the window stop meets the jamb so the paint doesn't pop off in large flakes as you pry the stop off. Removing one stop is sufficient to remove the bottom sash.

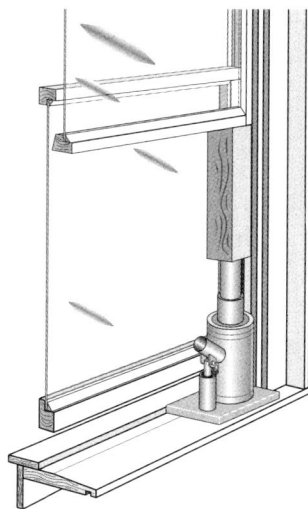

Lifting an upper window sash: First cut paint away from around the sash inside and outside. Then lift with leverage or a jack.

- ✓ Scrape excess paint from the sashes and the window sill. You may need to plane the sides so the window operates smoothly.

- ✓ Apply vinyl V-strip to the side jambs, and bronze V-strip to the meeting rail on the top sash. The point of the bronze V goes skyward. The weatherstrip is caulked on its back side and stapled in place, as shown in the illustration.

5.4 WINDOW REPLACEMENT SPECIFICATIONS

The purpose of these specifications is to guide the selection and installation of replacement windows. Improper window installation can cause air leakage, sound leakage, and water leakage.

Existing window openings may have moisture damage and air leakage. These conditions require repair during the window replacement process.

Included here are specifications for two special window safety considerations.

1. Windows in high risk areas, such as around doors and walkways, must have safety glass.

2. Windows are part of fire escape planning for homes, so this egress function must be recognized and accommodated.

5.4.1 Window Replacement Options

Consider the following three approaches for replacing windows in homes with membrane drainage systems, which utilize a water resistive barrier (such as Tyvek or asphalt felt) behind the siding.

The existing condition of the windows and siding are the most important considerations for selecting one of the three options below.

1. Ideally, windows should be replaced during siding replacement so that the flashing can be integrated perfectly with the wall's water resistive barrier.

   a. Use this method when moisture damage to both siding and window is evident.

   b. Replace all moisture damaged components during window replacement, including framing members if necessary.

2. If the siding will not be replaced, and the existing window is moisture damaged or has other problems, the existing window can be completely removed to expose the rough opening.

   a. Use this method whenever moisture damage is observed to the existing window sill, jambs, or exterior trim.

   b. Replace all moisture damaged components during window replacement, including nearby framing and siding.

c. Flash the rough opening and integrate this flashing with the home's water resistive barrier.

3. If cost is the primary consideration, a replacement window may be installed within the existing window jambs and sill when the existing window frame is undamaged by moisture and well integrated into the building exterior.

    a. The manufacturer of the window should approve this replacement process and their instructions should be carefully observed.

    b. Window jambs and sill must show no signs of deterioration from moisture.

The option chosen for replacement window installation also depends on the following site considerations.

- ✓ Annual rainfall
- ✓ Window orientation with reference to wind, rain, and sun
- ✓ The distance of the window's head jamb from a horizontal protective roof overhang

The more exposed the window is to weather, the more effort is merited to optimize its watertight performance.

5.4.2 Window Energy Specifications

Installing new windows incurs a large labor expense so they should be as energy-efficient as budget allows.

1. Replacement windows must have a U-factor less than or equal to U-0.32. Lower is better, especially in cold climates.

2. Replacement windows facing east or west in air conditioned homes should have a solar heat-gain coefficient (SHGC) of less than or equal to 0.35. Lower is better, especially in hot climates.

NFRC label: The key selection criteria for window shopping is displayed on the NFRC label.

3. Consider advanced window designs with triple glazing, insulated frames and sashes, gas fillings between the glass panes, and less conductive spacers between the glass panes.

Window Accessory Specifications

Accessories are extra components used by installers to attach the window, seal the window into the home's water resistive barrier, and complete the window installation.

1. Windows must be shimmed as necessary using flat shims of the correct thickness to give the proper support at the sill.

2. All fasteners used for window installation must be corrosion resistant, according to ASTM B 633, B 766 or B456.

3. Flashing for new windows must be reinforced, coated flexible flashing, designed for exterior water protection.

4. Sealants should be compatible with the materials they seal.

5. In joints more than $^3/_{16}$ inch in width, backer rod must be used to control sealant depth and prevent three-sided adhesion by the caulking.

6. All sides of exterior wood, used in the installation, must be primed with exterior primer or other equivalent wood sealer.

7. Metal window components must be protected from dissimilar metals or corrosive materials.

5.4.3 Removing Old Windows

Existing windows should be removed without damaging the home's interior finish, siding, exterior trim, and the water resistive barrier.

1. Protect the interior of the home from construction debris.

2. Remove window sashes, jambs, and/or siding, depending on the window-replacement method chosen.

3. Repair moisture damage to the rough opening before installing the new window.

5.4.4 Installing Replacement Windows

The most important considerations for installing new windows is that the window installation is weathertight and airtight.

Water leakage is a serious concern because it deteriorates building components around the window. To prevent water leakage in frame buildings, the window must be carefully integrated into the home's water resistive barrier (WRB). A frame home's WRB is a waterproof breathable membrane that stops rain water penetration through the siding from dampening the sheathing underneath. The integration of the new window and existing WRB is achieved using special tapes or adhesive flexible flashing.

Installing Windows Within the Rough Opening

Remove the existing window, window jambs, sill, and exterior trim. If the entire house is not re-sided at the same time, the nearby siding may be removed to allow proper installation of flashing. Then install the replacement window in the rough opening in a similar manner as new construction.

1. Use whatever shimming assembly is necessary including flat shims, a shimmed flat sill, or a sill pan to support the replacement window on a solid, level, and water-resistant sill surface. The window's weight should not be supported by the flanges.

2. Flash the new window around its perimeter with approved flashing. Install the flashing from bottom to top (like roof shingles) so water cannot enter the wall.

Building paper is
wrapped into the rough opening. Flashing
is installed only at the top of the opening.

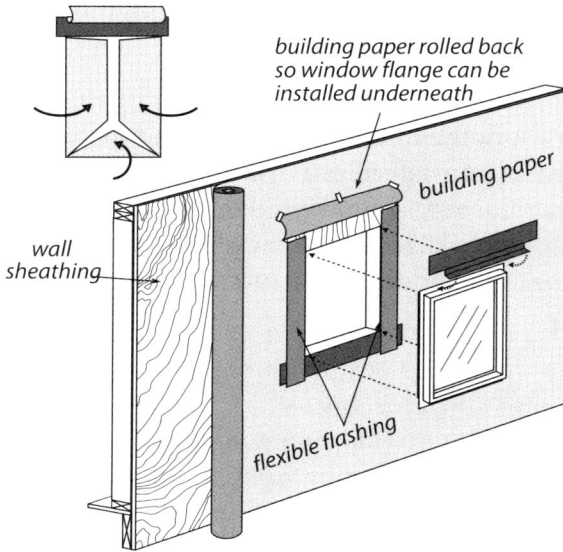

*building paper rolled back
so window flange can be
installed underneath*

building paper

*wall
sheathing*

**Preferred method to
flash a flanged
window:** To prevent
water damage, the
flange must be
flashed into the water
resistive barrier of the
home. Water that
penetrates the siding
is stopped by building
paper that is correctly
installed.

flexible flashing

3. At the top of the window, fit the window's flange **underneath** the home's water resistive barrier. At the sides
and bottom, fit the flange **between** the siding and the
water resistive barrier.

4. Install caulking or butyl putty tape on the window
flange before installing the window. Follow the manufacturer's recommendations on sealant and its application.

5. When using caulking as the window sealant, the window must be installed immediately after caulking application before the caulking becomes contaminated or
forms a skin.

6. Use fasteners with heads wide enough in diameter to
span the holes or slots in the window flange.

7. Avoid over-driving the fasteners or otherwise deform-
ing the window flange.

8. air seal the space between the window frame and the rough opening, or between the old window jambs and sill. Use one-part foam or foam backer rod with caulking. Merely stuffing this gap with fiberglass insulation does not create an effective air seal.

9. Windows that are exposed to wind-driven rain or without overhangs above them should have a rigid cap flashing to prevent rainwater from draining onto the window. The cap flashing should overlap the sides of the window enough to divert water away from vertical joints bordering the window, exterior window trim, and siding.

Installing Replacement Windows: Existing Jambs and Sill

When the contractor and homeowner are confident about the condition of the existing rough framing, jambs, and sill, replacement windows are often installed without stripping the assembly out to the rough frame. Air sealing will be more difficult and probably less effective than when the entire assembly is re-built.

1. Install flat shims to provide a level surface to support under the vertical structural members of the new window frame. The window should not be supported by the flanges.

2. Protect the existing sill with metal and plastic flashing if necessary for drainage and to protect the protruding wood sill.

Block frame window installation:
When installing a replacement window in the existing window frame or in a masonry wall the Block frame method is common. Windows installed Block frame have no flange. Install the window against a stop from inside the building. Caulk the stop and foam between the new window frame and the old one.

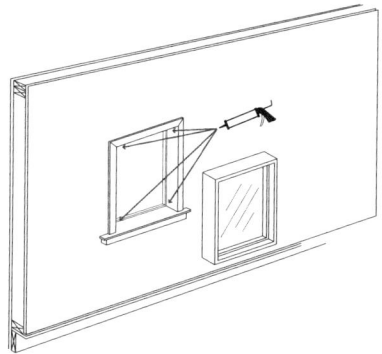

3. Seal the replacement window to a continuous stop during installation.

4. Use correctly sized shims where the side jambs of the replacement window are fastened into the side jambs of the old window.

5. Seal the space between the new window frame and old window frame with one-part foam.

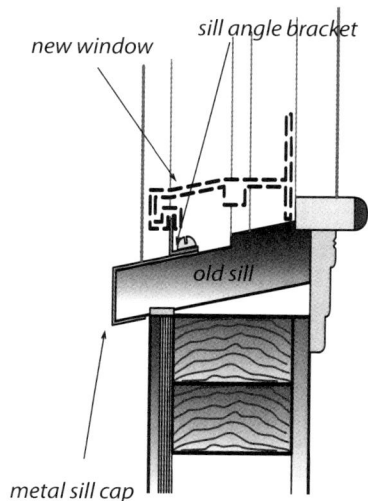

new window *sill angle bracket*

old sill

metal sill cap

Replacement window and sill angle:
The sill angle supports the window level and the sill cap covers and protects the aged sill.

5.5 WINDOW SAFETY SPECIFICATIONS

Windows have special requirements for breakage-resistance in areas that are prone to glass breakage, and for fire escape in bedrooms. This safety information is included here because of the difficulty of obtaining it elsewhere.

5.5.1 Windows Requiring Safety Glass

Safety glass must be either laminated glass or tempered glass bearing a permanent label identifying it as safety glass.

Instead of safety glazing, glazed panels may have a protective bar installed on the accessible sides of the glazing 34 to 38 inches above the floor. The bar shall be capable of withstanding a horizontal load of 50 pounds per linear foot without contacting the glass and be a minimum of $1^1/_2$ inches in width.

Safety glass or a protective bar is required in the following conditions.

✓ Glazing wider than 3 inches in entrance doors.

✓ Glazing in fixed and sliding panels of sliding doors and panels in swinging doors other than wardrobe doors.

✓ Glazing in fixed or operable panels adjacent to a door where the nearest exposed edge of the glazing is within a 24-inch arc of the vertical edge of the door in a closed position and where the bottom edge of the glazing is less than 60-inches above the floor or walking surface unless there is an intervening wall or permanent barrier between the door and the glazing.

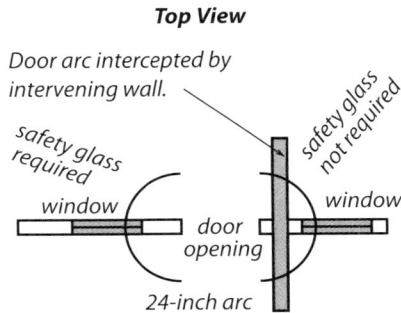

Top View

Door arc intercepted by intervening wall.

safety glass required

safety glass not required

window

window

door opening

24-inch arc

Safety glass around doors: A window near a door must be glazed with safety glass when the window is less then 24 inches from the door and less than 60 inches from the floor.

✓ Glazing in an individual fixed or operable panel that meets all of the following conditions:

1. An exposed area of an individual pane greater than 9 square feet, and

2. An exposed bottom edge less than 18 inches above the floor.

3. An exposed top edge greater than 36 inches above the floor.

4. One or more walkways within 36 inches horizontally of the glazing.

5. Glazing in any portion of a building wall enclosing showers, hot tubs, whirlpools, saunas, steam rooms, and bathtubs where the bottom exposed edge is less than 5 feet above a standing surface or drain inlet.

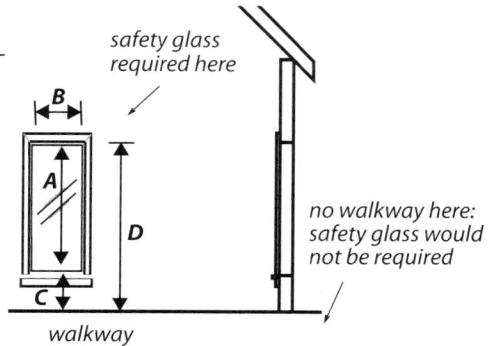

safety glass required here

no walkway here: safety glass would not be required

walkway

A x B = greater than 9 ft², and
C = less than 18 inches from floor, and
D = more than 36 inches from floor.

Large windows near walking surfaces:
Safety glass is required in picture windows with a nearby walkway.

5.5.2 Fire Egress Windows

Windows are the designated fire escape for many homes and should offer a minimum opening for a person's escape from a fire. The following specifications must be observed when replacing windows, regardless of the compliance of existing windows with fire-egress specifications.

1. Windows installed in bedrooms must observe the specifications for egress windows described here.

2. Each bedroom must have one egress window.

3. Egress windows must provide an opening that is at least 20 inches wide and at least 24 inches high.

4. Egress windows must provide an opening with a clear area of at least 5.7 square feet.

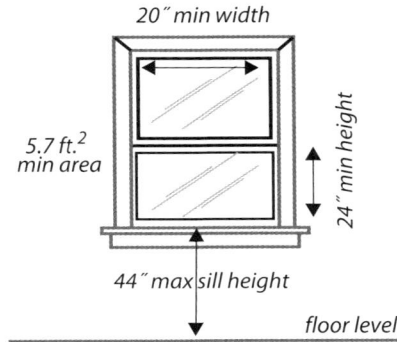

Egress windows: Windows for fire escape must be large enough and a convenient distance from the floor.

5. The finished sill of the egress window must be no higher off the floor than 44 inches.

5.6 DOOR REPLACEMENT AND REPAIR

Doors suffer a lot of wear because of their jobs as portals for buildings. Doors need repair when they are damaged and replacement when repair costs exceed the cost of a new insulated door.

Install flashing around doorways according to the specifications in *"Installing Replacement Windows" on page 174.*

5.6.1 Door Replacement

Exterior doors should be replaced with exterior-grade foam core doors. New doors should be painted or sealed to prevent deterioration.

5.6.2 Door Repair

Doors have a small surface area and their air leakage is more of a localized comfort problem than a significant energy problem most of the time. However, door operation affects building security and durability, so doors are often an important repair priority.

Door repair improves home security and building durability. Door repair can also save energy if the door currently has a poor fit.

Door weatherstrip, thresholds and sweeps are marginally cost effective. These measures may be used only if they are found to be cost effective.

- ✓ Tighten door hardware and adjust stops so door closes snugly against its stops.

- ✓ Use a durable stop-mounted or jamb-mounted weatherstrip material to weatherstrip the door. New weatherstrip must form a tight seal with no buckling or gaps when installed.

- ✓ Plane or adjust the door so it closes without rubbing or binding on the stops and jambs, especially in homes that may have lead paint.

- ✓ Install thresholds and door sweeps if needed to prevent air leakage. These air seals should not bind the door. Thresholds should be caulked on both sides at the sill and jamb junction.

- ✓ Replace missing or inoperable lock sets.

- ✓ Reposition the lock set and strike plate.

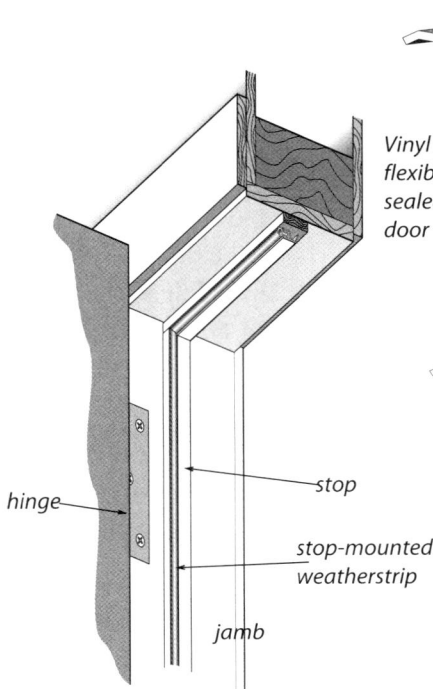

Vinyl flap weatherstrip is particularly flexible, allowing the door to remained sealed with seasonal movements of the door

Silicone bulb weatherstrip is much more flexible than vinyl bulb and therefore seals better.

Bronze v-strip mounts on the door jamb and is very durable.

Labels on diagram: hinge, stop, stop-mounted weatherstrip, jamb

Weatherstripping doors: Weatherstripping doors is mainly a comfort retrofit. The door should be repaired before weatherstripping by tightening hinges and latches. The door stop should fit tightly against the door when it is closed.

✓ Reposition stops if necessary.

✓ Seal gaps between the stop and jamb with caulk.

✓ Install a door shoe if needed to repair damage.

If the door binds at the top, check the tightness of screws in the top hinge and tighten them if necessary. If the hinges are tight, check the space between the door and the frame's hinge-side. If there's a $1/4$-inch gap between door and frame on the hinge side, you can give $1/8$ inch of that gap to the latch side by deepening the mortise – the chiseled-out section of door frame directly under the hinge. If the door is too tight to the hinge side, install one or more pieces of cardboard underneath a hinge.

Doors can be adjusted by moving the hinges in or out. Moving the top hinge in moves the door upward and toward the hinge. Moving the top hinge out drops the door down and moves the door away from the hinge.

If a door won't latch, inspect the door stops and weatherstripping to see if they are binding. If there is no obvious problem with the weather-strip or stops, move the strike plate out slightly or use a file to remove a little metal from the strike plate. The strike plate is mortised into the door frame that receives the latch. Use toothpicks or dowels to patch widened screw holes if you have to move the strike plate.

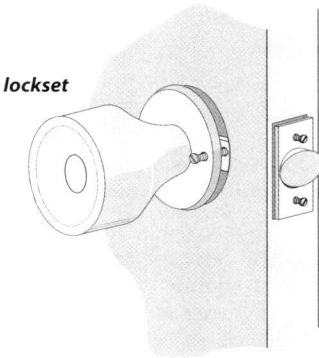

lockset

strike plate

Minor door repair: Tightening and adjusting locksets, strike plates, and hinges helps doors work better and seal tighter.

Chapter 6:   Mobile Homes

Mobile homes typically use more energy per square foot than site-built homes. Fortunately, their consistent construction makes them easy to weatherize. Mobile homes aren't governed by the International Residential Code 2009, allowing the contractor more flexibility in treating mobile homes' unique problems.

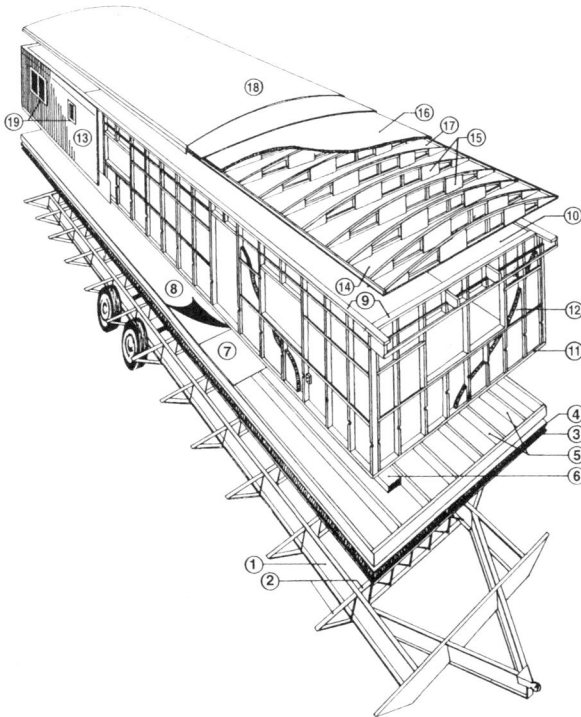

Typical Components of a Mobile Home: 1–Steel chassis. 2–Steel outriggers and cross members. 3–Underbelly. 4–Fiberglass insulation. 5–Floor joists. 6–Heating/air conditioning duct. 7–Decking. 8–Floor covering. 9–Top plate. 10–Interior paneling. 11–Bottom plate. 12–Fiberglass insulation. 13–Metal siding. 14–Ceiling board. 15–Bowstring trusses. 16–Fiberglass insulation. 17–Vapor barrier. 18–Galvanized steel one-piece roof. 19–Metal windows.

Insulation upgrades save the most energy in mobile homes, though sealing shell and duct air leaks presents good opportunities, too. Mobile home heating replacement is often cost-effective when a customer's energy usage is high.

6.1 MOBILE HOME AIR SEALING

The location and relative importance of mobile home air leaks was a mystery before blower doors. Some mobile homes are fairly airtight, and others are very leaky. Air leakage serves as ventilation in most mobile homes. Observe the minimum ventilation requirements outlined in *"Evaluating Home Ventilation Levels" on page 34*.

A duct airtightness tester, which pressurizes the ducts is the best way to measure and evaluate duct air sealing. For evaluating mobile home duct tightness, the blower door used in conjunction with a pressure pan does a good job of detecting air leaks. *See "Pressure Pan Testing" on page 58.*

Most mobile home duct sealing is performed through the belly. This work is more difficult once the belly has be re-insulated. Inspect the ductwork and seal any major leaks, such as disconnected trunk lines, before insulating the belly.

Table 6-1: air sealing Locations & Typical CFM$_{50}$ Reduction

air sealing Procedure	Typical CFM$_{50}$ Reduction
Patching large air leaks in the floor, walls and ceiling	200–900
Sealing floor as return-air plenum	300–900
Sealing leaky water-heater closet	200–600
Sealing leaky supply ducts	100-500
Installing tight interior storm windows	100-250
Caulking and weatherstripping	50–150

Mobile home shell air leakage is often substantially reduced when insulation is installed in roofs, walls, and belly cavities. Prioritize your efforts by performing these tasks in this order.

1. Assess the insulation levels. If adding insulation is cost-effective, perform the usual pre-insulation air sealing measures that also prevent spillage of insulation out of the cavity.

2. Install cavity insulation. Perform duct sealing first if the belly is to be insulated.

3. Re-check the air leakage rate.

4. Perform additional air sealing as needed.

6.1.1 Shell Air Leakage Locations

Blower doors have pointed out the following shell locations as the most serious Air Leakage sites.

✓ Plumbing penetrations in floors, walls, and ceilings. Water-heater closets with exterior doors are particularly serious Air Leakage problems, having large openings into the bathroom and other areas

✓ Torn or missing underbelly, exposing flaws in the floor to the ventilated crawl space

✓ Large gaps around furnace and water heater chimneys

✓ Severely deteriorated floors in water heater compartments

✓ Gaps around the electrical service panel box, light fixtures, and fans

✓ Joints between the halves of double-wide mobile homes and between the main dwelling and additions

Note: Window and door air leakage is more of a comfort problem than a serious energy problem.

6.1.2 Duct Leak Locations

Blower doors and duct testers have pointed out the following duct locations as the most serious energy problems.

- ✓ Floor and ceiling cavities used as return-air plenums — These floor return systems should be eliminated and replaced with return-air through the hall or a large grille in the furnace-closet door.

- ✓ Joints between the furnace and the main duct — The main duct may need to be cut open from underneath to access and seal these leaks between the furnace, duct connector, and main duct. With electric furnaces you can access the duct connector by removing the resistance elements. For furnaces with empty A-coil compartments, you can simply remove the access panel to seal the duct connector.

- ✓ Joints between the main duct and the short duct sections joining the main duct to a floor register

- ✓ Joints between register boots and floor

- ✓ The poorly sealed ends of the duct trunk, which often extend beyond the last supply register

- ✓ Disconnected, damaged or poorly joined crossover ducts

- ✓ Supply and return ducts for outdoor air conditioner units

- ✓ Holes cut in floors by tradesmen.

- ✓ New ductwork added to supply heat to room additions

Be sure to seal floor penetrations and ductwork before performing any belly repair. Pollutants in the crawl space such as mold, mildew, and fiberglass will be disturbed by repair work and can be drawn into the home by duct depressurization.

6.1.3 Belly Pressure Test

✓ Mobile home supply duct leaks pressurize the belly cavity. Follow these steps to perform this rough test to determine if duct leaks are present and their general location.

✓ Repair the rodent barrier.

✓ Turn on the air handler.

✓ Insert a manometer hose into the belly through the rodent barrier and test the pressure with-reference-to the outdoors.

✓ Start near the furnace, and work your way toward the ends alongside the trunk line. A pressure rise gives you a rough idea of the location of leaks, size of leaks, and tightness of the nearby rodent barrier.

✓ Repair the ducts and re-test.

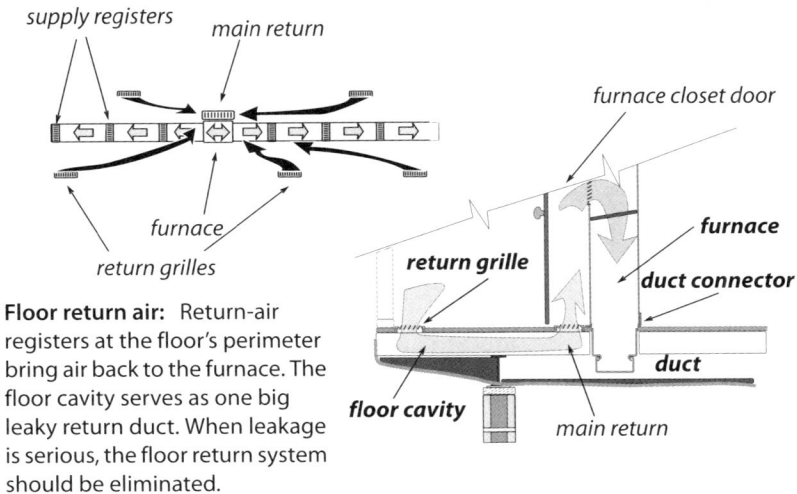

supply registers   main return

furnace closet door

furnace

duct connector

return grille

duct

furnace

return grilles

floor cavity

main return

Floor return air: Return-air registers at the floor's perimeter bring air back to the furnace. The floor cavity serves as one big leaky return duct. When leakage is serious, the floor return system should be eliminated.

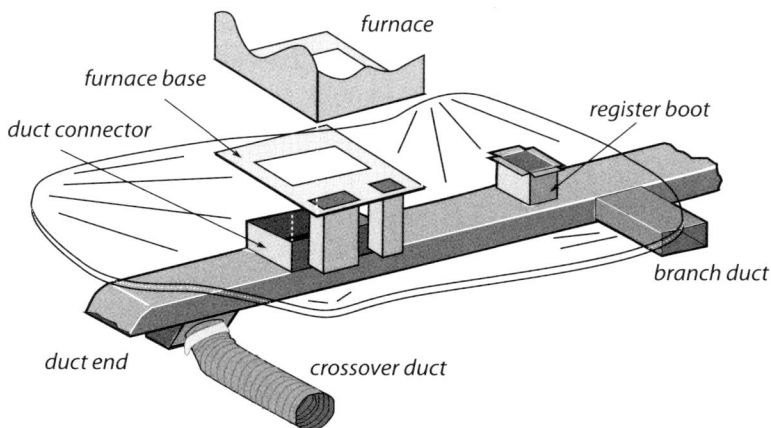

furnace

furnace base

register boot

duct connector

branch duct

duct end

crossover duct

Mobile home ducts: Mobile home ducts leak at their ends and joints – especially at the joints beneath the furnace. The furnace base attaches the furnace to the duct connector. Leaks occur where the duct connector meets the main duct and where it meets the furnace. Branch ducts are rare, but easy to find, because their supply register isn't in line with the others. Crossover ducts are found only in double-wide and triple-wide homes.

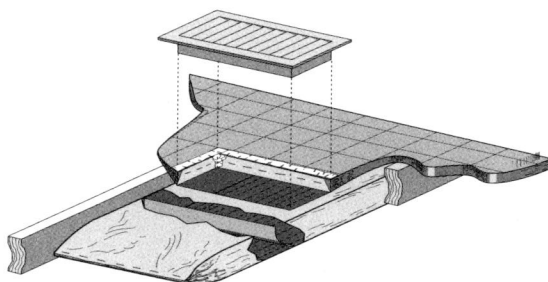

Sealing the end of the main duct: The main duct is usually capped or crimped loosely at each end, creating a major air leakage point. Seal this area and improve airflow by installing a sheet metal ramp, accessed through the last register, inside the duct. Seal the ramp to the ductwork with metal tape and silicone or mastic.

6.2 Mobile Home Insulation

Address all significant moisture problems before insulating. The most important single moisture-control measure is installing a ground-moisture barrier. *See "Moisture Problems" on page 219. See also "Ground Moisture Barriers" on page 141.*

6.2.1 Insulating Mobile Home Roof Cavities

Blowing a closed mobile home roof cavity is similar to blowing a closed wall cavity, only the insulation doesn't have to be as dense. Use fiberglass blowing wool because cellulose is too heavy and absorbs water too readily for use around a mobile home's lightweight sheeting materials.

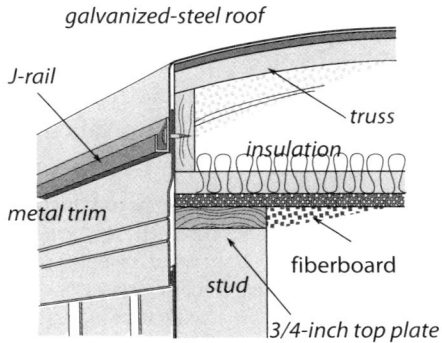

Bowstring roof details: Hundreds of thousands of older mobile homes were constructed with these general construction details.

There are three common and effective methods for blowing mobile home roof cavities.

1. Cutting a square hole in the metal roof and blowing fiberglass through a flexible fill-tube.

2. Disconnecting the metal roof at its edge and blowing fiberglass through a rigid fill-tube.

3. Blowing fiberglass through holes drilled in the ceiling.

Preparing to Blow a Mobile Home Roof

Perform these steps before insulating mobile home roofs.

- ✓ Reinforce weak areas in the ceiling.

- ✓ Inspect the ceiling and seal all penetrations.

- ✓ Take steps to maintain safe clearances between insulation and recessed light fixtures and ceiling fans.

Blowing Through the Top

Blowing through the roof top does a good job of filling the critical edge area with insulation, and the patches are easy to install if you have the right materials. It is important to complete the work during good weather, however, since the roof will be vulnerable to rain or snow during the job.

If the roof contains a strongback running the length of the roof, the holes should be centered over the strongback, which is usually near the center of the roof's width. A strongback is a 1-by-4 or a 1-by-6, installed at a right angle to the trusses near their center point, that adds strength to the roof structure.

1. Cut 10-inch square holes at the roof's apex on top of every second truss. Each square hole permits access to two truss cavities.

2. Use a 2-inch or 2-1/$_2$-inch diameter fill-tube. Insert the fill-tube and push it forcefully out toward the edge of the cavity.

3. Blow fiberglass insulation into each cavity.

Roof-top insulation: Blowing fiberglass insulation through the roof top is effective at achieving good coverage and density on almost any metal roof.

4. Stuff the area under each square hole with a piece of unfaced fiberglass batt

so that the finished roof patch will stand a little higher than the surrounding roof.

5. Patch the hole with a 14-inch-square piece of stiff galvanized steel, sealed with roof cement and screwed into the existing metal roof.

6. Cover the first patch with a second patch, consisting of an 18-inch-square piece of foil-faced butyl rubber.

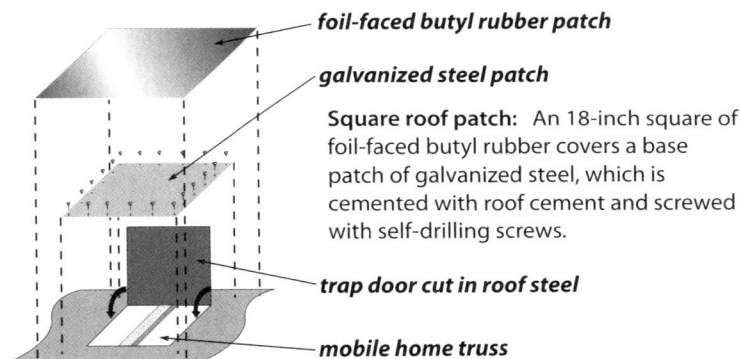

foil-faced butyl rubber patch

galvanized steel patch

Square roof patch: An 18-inch square of foil-faced butyl rubber covers a base patch of galvanized steel, which is cemented with roof cement and screwed with self-drilling screws.

trap door cut in roof steel

mobile home truss

Blowing a Mobile Home Roof from the Edge

Erect scaffold to be performed this procedure safely and efficiently. Mobile home metal roofs are usually fastened only at the edge, where the roof joins the wall.

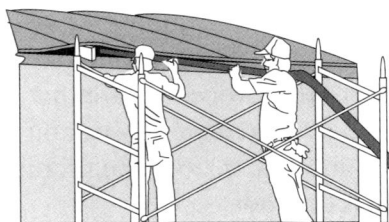

1. Remove the screws from the metal j-rail at the roof edge. Also remove staples or other fasteners, and scrape off putty tape.

Roof-edge blowing: Use a rigid fill tube to blow insulation through the roof edge. This avoids making holes in the roof itself, though this process requires much care in refastening the roof edge.

2. Pry the metal roof up far enough to insert a 2-inch-diameter, 10- to 14-foot-long rigid fill-tube. Two com-

mon choices are steel muffler pipe and aluminum irrigation pipe. Inspect the cavity with a bright light to identify any wires or piping that could be damaged by the fill tube.

3. Blow insulation through the fill-tube into the cavity. Turn off the insulation-material feed and blower on the blowing machine when the tube is a couple feet from the roof edge, in order to avoid blowing insulation out through the opening in the roof edge. Stuff the last foot or two with unfaced fiberglass batts.

4. Fasten the roof edge back to the wall using galvanized roofing nails, a new metal j-rail, new putty tape, and larger screws. The ideal way to re-fasten the metal roof edge is with air-driven galvanized staples, which is the way most roof edges were attached originally.

The re-installation of the roof edge is the most important part of this procedure. Putty tape must be replaced and installed as it was originally. This usually involves installing a layer of putty tape or a bead of high quality caulk under the metal roof and another between the metal roof edge and the j-rail.

The advantages of blowing through the edge is that if you have the right tools, including a powered stapler. This method can be very fast and doesn't require cutting into the roof. The disadvantages of this procedure are that you need scaffolding to work at the edges, and it won't work on roof systems with a central strongback that stops the fill tube from reaching all the way across the roof.

Blowing a Mobile Home Roof from Indoors

The advantage to this method is that you are indoors, out of the weather. The disadvantages include being indoors where you can make a mess — or worse, damage something.

Blowing the roof cavity from indoors requires the drilling of straight rows of 3-inch or 4-inch holes and blowing insulation into the roof cavity through a fill tube.

Follow this procedure.

1. Drill a 3-inch or 4-inch hole in an unseen location to discover whether the roof structure contains a strongback that would prevent blowing the roof cavity from a single row of holes.

2. Devise a way to drill a straight row of holes down the center of the ceiling. If a strongback exists, drill two rows of holes at the quarter points of the width of the ceiling.

3. Insert a flexible plastic fill tube into the cavity, and push it as far as possible toward the edge of the roof.

4. Fill the cavity with tightly packed fiberglass insulation.

5. Cap the holes with manufactured plastic caps. Care must be taken not to damage the holes so that the plastic hole covers fit properly. You can also install a piece of painted wood trim over the line of holes.

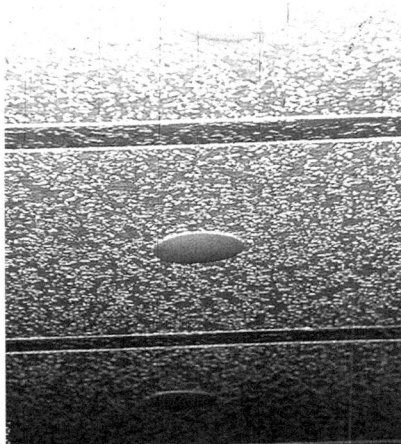

Blowing through the ceiling: The technician pushes the fill-tube into the cavity and out near the edge of the roof. The holes are drilled in a straight line for appearance sake.

6.2.2 Mobile Home Sidewall Insulation

The sidewalls of many mobile homes are not completely filled with insulation. This reduces the nominal R-value of the existing wall insulation because of convection currents and air leakage. Consider the following steps for adding insulation to partially filled mobile home walls.

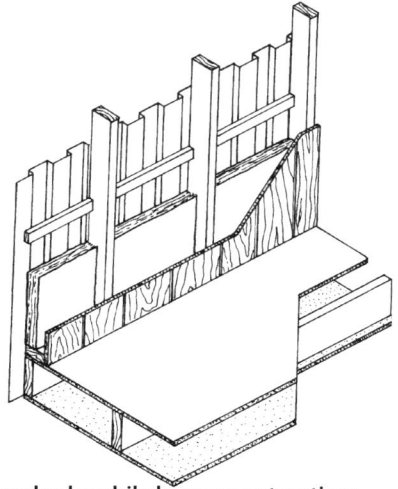

Standard mobile home construction: 2-by-4 walls and 2-by-6 floor joists are the most common construction details.

1. Check the interior paneling and trim to make sure they are securely fastened to the wall. Repair holes in interior paneling and caulk cracks at seams to prevent indoor air from entering the wall. Note the location of electrical boxes and wire to avoid hitting them when you push the fill tube up the wall.

2. Remove the bottom horizontal row of screws from the exterior siding. If the vertical joints in the siding interlock, fasten the bottom of the joints together with $^1/_2$-inch sheet metal screws to prevent the joints from

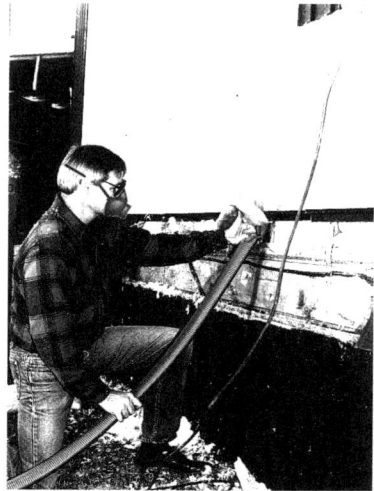

Adding insulation to mobile home walls: A technician uses a fill tube to install more insulation in a partially filled mobile home wall.l.

coming apart. Pull the siding and existing insulation away from the studs, and insert the fill tube into the cavity with the point of its tip against the interior paneling.

3. Push the fill tube up into the wall cavity until it hits the top plate of the wall. The tube should go in to the wall cavity 7-to-8 feet. It is important to insert the tube so that its natural curvature presses its tip against the interior paneling. When the tip of the fill tube, cut at an angle, is pressed against the smooth paneling, it is least likely to snag the existing insulation on its way up the wall. If the fill tube hits a belt rail or other obstruction, twisting the tube will help its tip get past the obstruction.

4. Stuff a piece of fiberglass batt into the bottom of the wall cavity around the tube to prevent insulation from blowing out of the wall cavity. Leave the batt in-place at the bottom of the wall, when you pull the fill tube out of the cavity. This piece of batt acts as temporary gasket for the hose and insulates the very bottom of the cavity after the hose is removed. This batt also eliminates the need to blow insulation all the way to the bottom, preventing possible spillage and overfilling. If you happen to overfill the bottom of the cavity, reach up inside the wall to pack or remove some insulation, particularly any that lies between the loose siding and studs.

5. Draw the tube down and out of the cavity about 6 inches at a time. Listen for the blower fan to indicate strain from back-pressure in the wall. Watch for the insulation to slow its flow rate through the blower hose at the same time. Also watch for slight bulging of the exterior siding. These signs tell the installer when to pull the tube down.

6. Carefully refasten the siding using the same holes. Use screws that are slightly longer and thicker than the original screws.

6.2.3 Mobile Home Floor Insulation

Mobile home floor insulation is a good energy-saving measure in cool climates. The original insulation is usually fastened to the bottom of the floor joists, leaving much of the cavity uninsulated and subject to convection currents. This greatly reduces the insulation's R-value. Blown-in belly insulation also tends to control duct leakage.

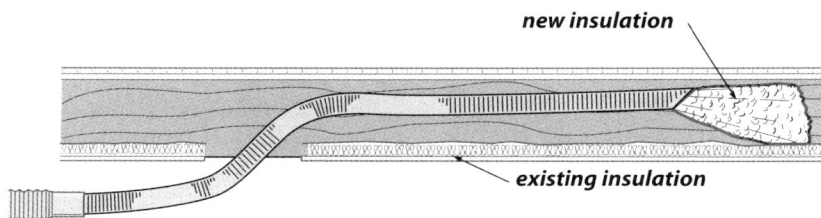

Blowing bellies: A flexible fill-tube, which is significantly stiffer than the blower hose, blows fiberglass insulation through a hole in the belly from underneath the home.

Preparing for Mobile Home Floor Insulation

Prior to installing floor insulation, always perform these repairs.

- ✓ Repair plumbing leaks.
- ✓ Tightly seal all holes in the floor.
- ✓ Inspect and seal ducts.
- ✓ Repair the rodent barrier.
- ✓ Install a ground-moisture barrier in the crawl space if the site is wet.

Insulating the Floor

Two methods of insulating mobile home floors are common. Blown fiberglass is recommended over cellulose for either method.

1. Drilling through the 2-by-6 rim joist and blowing fiberglass through a rigid fill tube into the belly.

2. Blowing fiberglass insulation through a flexible fill tube or a rigid fill tube into the underbelly.

First repair all holes in the belly. Use mobile home belly-paper, silicone sealant, and stitch staples. Use these same patches over the holes cut for fill-tubes. Screw wood lath over weak areas if needed.

When blowing through holes from underneath the home, consider blowing through damaged areas before patching them.

Blowing a floor through the belly: The technician inserts a rigid fill tube through the belly to blow insulation into the floor cavity and underbelly.

Identify any plumbing lines, and avoid installing insulation between them and the living space if freezing could be an issue. This may require running a piece of belly-paper under the pipes, and insulating the resulting cavity, to include them in the heated envelope of the home.

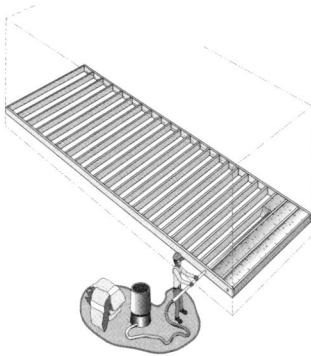

Blowing crosswise cavities: Blowing insulation into belly is easy if the floor joists run crosswise. However, the dropped belly requires more insulation than a home with lengthwise joists.

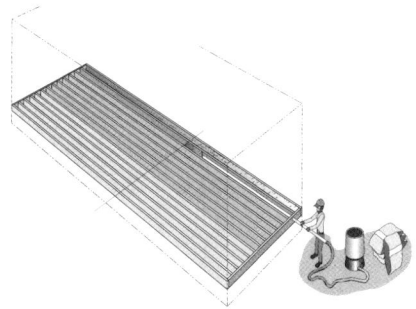

Blowing lengthwise cavities: Floors with lengthwise joists can rarely be filled completely from the ends because of the long tubing needed. The middle can be filled from underneath.

Unfaced fiberglass batts may also be used to insulate floor sections where the insulation and belly are missing. The insulation should be supported by lath, twine, or insulation supports. This is a good approach when it is not cost-effective to insulate the entire belly.

6.3 MOBILE HOME WINDOWS AND DOORS

Repairing or replacing mobile home windows and doors is often part of a mobile home weatherization job. Installing storm windows or replacing existing windows is expensive per square foot and isn't as cost-effective as insulation. However, storm windows, window shading devices, and replacement windows are all energy conservation measures for mobile homes that are worth considering

6.3.1 Mobile Home Storm Windows

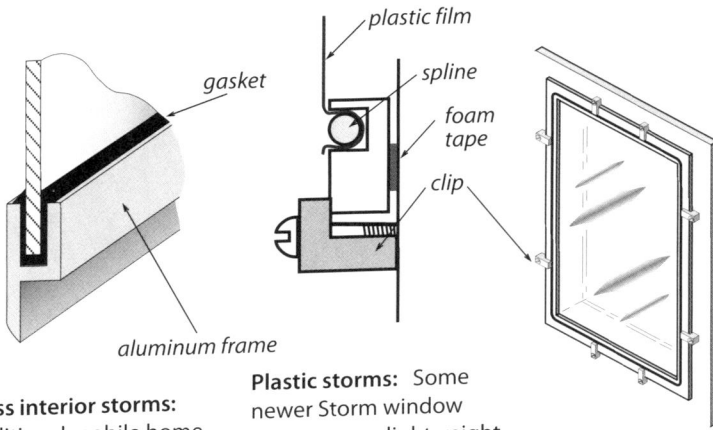

Glass interior storms:
Traditional mobile home storm windows have aluminum frames glazed with glass.

Plastic storms: Some newer Storm window designs use a lightweight aluminum frame and flexible or rigid plastic glazing.

Interior storm windows are common in mobile homes. These stationary interior storms serve awning and jalousie windows. Sliding interior storm windows pair with exterior sliding prime windows.

- ✓ Interior storm windows double the R-value of a single-pane window. They also reduce infiltration, especially in the case of leaky jalousie prime windows.

- ✓ Consider repairing existing storm windows rather than replacing them unless the existing storm windows cannot be re-glazed or repaired.

- ✓ When sliding primary windows are installed, use a sliding storm window that slides from the same side as the primary window. Sliding storm windows stay in place and aren't removed seasonally, and are therefore less likely to be lost or broken.

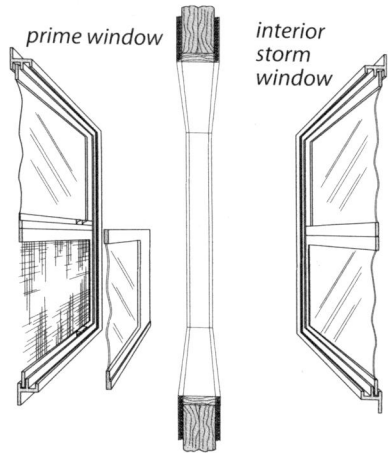

Mobile home double window: In mobile homes, the prime window is installed over the siding outdoors, and the storm window is installed indoors.

6.3.2 Replacing Mobile Home Windows

Replacement windows should have lower U-factors than the windows they are replacing. Inspect condition of rough opening members before replacing windows. Replace deteriorated, weak, or waterlogged framing members.

Prepare replacement window by lining the perimeter of the inner lip with $1/8$-inch thick putty tape. Caulk exterior window frame perimeter to wall after installing window.

6.3.3 Mobile Home Doors

Mobile home doors come in two basic types: the mobile home door and the house-type door. Mobile home doors swing outwardly, and house-type doors swing inwardly. House-type doors are available with pre-hung storm doors included.

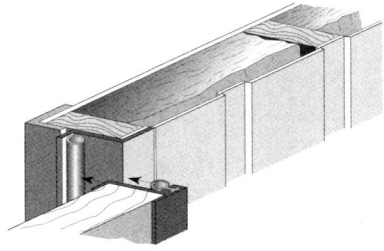

Mobile home door: Mobile home doors swing outwardly and have integral weatherstrip.

6.4 COOL ROOFS FOR MOBILE HOMES

Cool roof coatings reduce summer cooling costs and improve comfort by reflecting solar energy away from the home's roof and slowing the flow of heat into the home. They are shown to reduce overall cooling costs by 10-20%, and are a good choice for mobile homes or site-built homes with low slope or flat roofs. Cool roof coatings are usually bright white, and must have a reflectivity of at least 60% to meet the ENERGY STAR requirement for cool roof coatings.

Cool roof coatings are usually water-based acrylic elastomers, and are applied with a roller. They can be applied over most low-sloped roofing materials such as metal, built-up asphalt, bitumen, or single ply membranes. Some underlying materials require a primer to get proper adhesion-check the manufacturer's recommendations for asphalt-shingle roofs.

Surface preparation is critical when applying any coating. The underlying roofing materials must be clean so the coating will stick. Repairs should be performed if the existing roofing is cracked or blistered. Roof coating will not stick to dirty or greasy surfaces, and they cannot be used to repair roofs in poor conditions. Observe the following specifications when installing cool roof coatings.

✓ Install the coating when dry weather is predicted. Rain heavy dew, or freezing weather, if it happens within 24 hours of installation, will weaken the coating's bond to the underlying roofing.

✓ Protect any nearby windows, siding, or automobiles from splatters. For roller application, use a large brush for the edges, and a shaggy 1 to 1 1/2-inch roller on a 5- or 6- foot pole for the field. Run the coating up the roof jacks and other penetrations to help seal these areas. Install at least two coats, with second coat applied in the opposite direction to the firs to get more complete coverage. Allow a day for drying between coats.

✓ Clean the roof of debris and loose or detach roofing material.

✓ Wash the roof with a water/try-sodium phosphate (tsp) solution, or comparable mildew-cide, and scrub brush. Better yet, use a pressure washer.

✓ Buy the highest quality coatings, and look for those that are specifically formulated as mobile home roof coatings.

✓ Reinforce any open joints around skylights, pipe flashing, roof drains, wall transitions, or HVAC equipment. For build-up asphalt or bitumen roofs, repair any cracks, blisters, or de-laminations. Use polyester fabric and roof coating for these reinforcements and repairs by dipping fabric patches in the roof coating and spreading them over the existing roofing, or by laying dry fabric into a layer of wet coating. Smooth the patches down with a broad-knife or squeegee to remove bubbles or wrinkles. Allow any repairs to cure for 1 to 2 days before applying the topcoat.

✓ For metal roofs, sand any rusted areas down to sound metal. Install metal patches over any areas that are rusted through, followed by polyester patches as described above.

6.5 Mobile Home Skirting

The primary purpose of skirting is to keep animals out of the crawl space. Skirting must be vented to reduce moisture accumulation in many climates, so there isn't much value in insulating it.

Installation and repair of mobile home skirting is seldom cost-effective. Try to locate the thermal boundary at the floor of mobile homes.

Mobile Homes

Chapter 7: Health and Safety

This chapter introduces some of the most pressing hazards that your clients face in their homes, as well as those you face at work.

When you discover serious safety problems in a customer's home, you should inform the customer about the hazards and suggest how to eliminate them. Major hazards and potentially life-threatening conditions should be corrected before you begin work in the dwelling unless the you are making the corrections as part of their work.

House fires, carbon monoxide poisoning, moisture problems, and lead-paint poisoning are the most important health and safety problems that are related to building repair work.

✓ Inspect the home for fire hazards such as improperly installed electrical equipment, flammable materials stored near combustion appliances, or malfunctioning heating appliances. Discuss the problems with the client, and perform repairs if possible.

✓ Test combustion appliances for carbon monoxide production and other related hazards. Test the ambient air for carbon monoxide. Solve the problems causing these hazards.

✓ Find moisture problems and discuss them with the client. Never make moisture problems worse. *See page 219.*

✓ Follow the EPA Repair, Renovation, and Painting rules when working on homes built before 1978. Practice effective dust containment in all weatherization projects. *See page 231.*

7.1 Essential Combustion Safety Tests

The Building Performance Institute (BPI) requires that essential combustion safety tests be performed as part of all energy conservation jobs. BPI requires gas leak-testing and CO testing for all appliances. For naturally drafting appliances, either a worst-case venting test or zone-isolation test is also necessary. BPI considers naturally drafting appliances and venting systems to be obsolete for both efficiency and safety reasons. BPI strongly recommends that these obsolete appliances be replaced with modern sealed-combustion or power-vented combustion appliances.

7.1.1 Leak-Testing Gas Piping

Natural gas and propane piping systems may leak at their joints and valves. Find gas leaks with an electronic combustible-gas detector, often called a gas sniffer. A gas sniffer find all significant gas leaks if used carefully. Remember that natural gas rises from a leak and propane falls, so position the sensor accordingly.

Gas sniffer

- ✓ Sniff all valves and joints with the gas sniffer.

- ✓ Accurately locate leaks using a non-corrosive bubbling liquid, designed for finding gas leaks.

- ✓ All gas leaks must be repaired.

- ✓ Replace kinked or corroded flexible gas connectors.

- ✓ Replace flexible gas lines manufactured before 1973. The date is stamped on a date ring attached to the flexible gas line.

7.1.2 Carbon Monoxide (CO) Testing

CO testing is essential for evaluating combustion and venting. Measure CO in the vent of every combustion appliance you inspect and service. Measure CO in ambient air in both the home and combustion appliance zone (CAZ) during your inspection and testing of combustion appliances.

Vent Testing for CO

Testing for CO in the appliance vent is a part of combustion appliance tests. Vent testing is done under worst-case conditions. If CO is present in the undiluted combustion byproducts in a concentration of more than 100 parts per million (ppm measured) or 200 ppm (air-free), then the appliance fails the CO test.

Ambient Air Monitoring for CO

BPI standards require technicians to monitor CO during testing to ensure that air in the CAZ doesn't exceed 35 parts per million. If ambient CO levels in the combustion zone exceed 35 parts per million (ppm), stop testing for the your own safety. Ventilate the CAZ thoroughly before resuming combustion testing. Investigate indoor CO levels of 9 ppm or greater to find their cause.

Table 7-1: Testing Requirements for Combustion Appliances and Venting Systems

Appliance/Venting System	Required Testing
All direct-vent or power-vent combustion appliances	Gas leak test CO test at flue-gas exhaust outdoors Confirm venting system connected
Combustion appliances (with naturally drafting chimneys) in a mechanical room or attached garage supplied with outdoor combustion air and sealed from the home	Gas leak test CO test Confirm that CAZ is effectively air sealed from house and has combustion air from outdoors

Table 7-1: Testing Requirements for Combustion Appliances and Venting Systems

Appliance/Venting System	Required Testing
Naturally drafting chimney and appliance located within home	Gas leak test CO test Venting inspection Worst-case draft and depressurization testing

7.1.3 Worst-Case Testing for Atmospheric Venting Systems

Depressurization is the leading cause of backdrafting and flame roll-out in furnaces, boilers, and water heaters that vent into naturally drafting chimneys. The best option is to replace the older appliances and their naturally drafting venting systems with direct-vented or power-vented appliances with airtight venting systems. However, if the atmospheric appliances and venting systems must remain, perform the worst-case testing procedures documented here.

Worst-case vent testing uses the home's exhaust fans, air handler, and chimneys to create worst-case depressurization in the combustion-appliance zone (CAZ). The CAZ is an area containing one or more combustion appliances. During this worst-case testing, you can test for spillage, measure the indoor-outdoor pressure difference, and measure chimney draft.

Worst-case conditions do occur, and venting systems must exhaust combustion by-products even under these extreme conditions. Worst-case vent testing exposes whether or not the venting system exhausts the combustion gases when the combustion-zone pressure is as negative as you can make it. A sensitive digital manometer is the best tool for accurate and reliable readings of both combustion-zone depressurization and chimney draft.

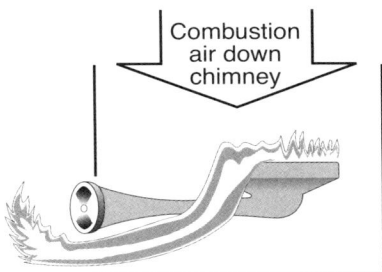

Flame roll-out: Flame roll-out, a serious fire hazard, can occur when the chimney is blocked, the combustion zone is depressurized, or during very cold weather before the appliance can establish draft.

Take all necessary steps to reduce spillage and strengthen draft as necessary based on testing.

A reading more negative than –5 pascals indicates a significant possibility of backdrafting.

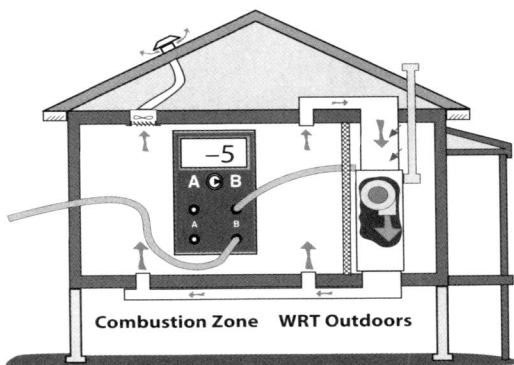

Worst-case depressurization: Worst-case testing is used to identify problems that weaken draft and restrict combustion air. The testing described here is intended to isolate the negative-pressure source.

7.1.4 Worst-Case Depressurization, Spillage, and CO

Worst-case testing creates a condition where all the possible factors, which could create spillage are activated. During the worst-case condition, combustion exhaust should continue to vent out the venting system. In preparation for worst-case testing, do the following.

1. Close all exterior doors, windows, and fireplace damper(s).

2. Set all combustion appliances to the pilot setting or turn them off at the service disconnect.

3. Measure and record the base pressure of the combustion appliance zone (CAZ) with reference to outdoors. If the digital manometer has a self-zeroing or "base" function, use this zeroing function to measure this baseline pressure.

Then establish worst-case conditions and measure the maximum worst-case depressurization, spillage, and CO.

1. Turn on the dryer and all exhaust fans.

2. Close the interior doors, which make the CAZ pressure more negative. Experiment by opening and closing interior doors while the air handler is operating.

3. Turn on the air handler, if present, using the "fan on" switch. Leave on if the pressure in the CAZ becomes more negative. Don't fire the burner of the combustion appliance yet.

4. Measure the net change in pressure from the CAZ to outside, correcting for the base pressure previously. Record the "worst-case depressurization" and compare to the table entitled, *Maximum CAZ or Mechanical Room Depressurization for Various Appliances" on page 214* for the tested appliance.

5. Next, fire the combustion appliances and test for spill-age and CO.

6. Fire the appliance with the smallest BTU capacity first and then the next largest and so on

7. Test for spillage at the draft diverter with a smoke gen-erator, a lit match, or a mirror. Note whether combus-tion by-products spill and how long after ignition that the spillage stops.

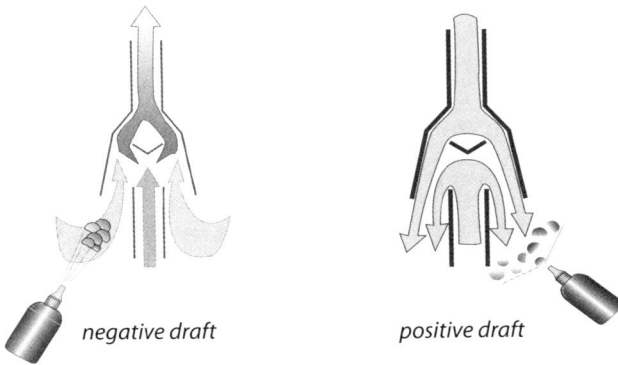

negative draft   positive draft

Negative Versus Positive Draft: With positive draft air flows down the chimney and out the draft diverter. A smoke bottle helps distinguish between positive and negative draft in atmospheric chimneys.

8. Test CO in the undiluted flue gases at 5 minutes.

9. If spillage in one or more appliances continues under worst-case for 1 minute or more, test the appliance again under natural conditions.

Table 7-2: Maximum CAZ or Mechanical Room Depressurization for Various Appliances

Appliance	Maximum Depressurization
Direct-vent appliance	50 pa (0.20 IWC)
Pellet stove with draft fan and sealed vent	15 pa (0.06 IWC)
Atmospherically vented oil and gas systems	5 pa (0.02 IWC)
Oil power burner and fan-assisted (induced-draft) gas*	
Closed controlled combustion	
Decorative wood-burning appliances	
Atmospherically vented water heater	2 pa (0.008 IWC)

*Individual fan-assisted (induced-draft) appliances with no vent hood attached to intact b-vent and oil appliances with flame retention head power burners are likely to vent safely at greater than 5 pascals depressurization but not enough test data is available to set a higher limit at this time. Since the appliances are possibly connected to an unsealed chimney and most spillage is through joints and the barometric damper these systems are included in the 5pa limit.

7.1.5 Improving Inadequate Draft

If measured draft is below minimum draft pressures, investigate the reason for the weak draft. Open a window, exterior door, or interior door to observe whether the addition of combustion air improves draft. If this added air strengthens draft, the problem usually is depressurization. If opening a window doesn't improve draft, inspect the chimney. The chimney could be blocked or excessively leaky.

Chimney Improvements to Solve Draft Problems

Suggest the following chimney improvements to solve draft problems uncovered during the previous testing.

- ✓ Remove chimney obstructions.
- ✓ Repair disconnections or leaks at joints and where the vent connector joins a masonry chimney.
- ✓ Measure the size of the vent connector and chimney and compare to vent-sizing information listed in Section 504 of the *International Fuel Gas Code*. A vent connector or chimney liner that is either too large or too small can result in poor draft.
- ✓ If wind is causing erratic draft, consider installing a wind-dampening chimney cap.
- ✓ If the masonry chimney is deteriorated, consider installing a new chimney liner.
- ✓ Increase the pitch of horizontal sections of vent.

Duct Improvements to Solve Draft Problems

Suggest the following duct improvements to solve draft problems uncovered during the previous testing.

- ✓ Seal all return-duct leaks near furnace, especially duct-to-furnace connection and filter-slot covers.

✓ air seal the CAZ from negative pressures caused by return registers. There should be no return registers in the CAZ.

✓ Improve balance between supply and return air by installing new return ducts, transfer grilles, or jumper ducts.

✓ Reducing depressurization from exhaust appliances

Consider suggesting the following remedies to depressurization caused by the home's exhaust appliances.

✓ Isolate furnace from exhaust fans and clothes dryers by air sealing between the CAZ and zones containing these exhaust devices.

✓ Reduce capacity of large exhaust fans.

✓ Provide make-up air for dryers and exhaust fans and/or provide combustion-air inlet(s) to combustion zone.

Table 7-3: Draft Problems and Solutions

Problem	Possible Solutions
Adequate draft never established	Remove chimney blockage, seal chimney air leaks, or provide additional combustion air as necessary.
Blower activation weakens draft	Seal leaks in the furnace and in nearby return ducts. Isolate the furnace from nearby return registers.
Exhaust fans weaken draft	Provide make-up or combustion air if opening a door or window to outdoors strengthens draft during testing.
Closing interior doors during blower operation weakens draft	Add return ducts, jumper ducts, or grilles between rooms.

7.1.6 Zone Isolation Testing for Atmospherically Vented Appliances

An isolated CAZ improves the safety of atmospherically vented appliances. The CAZ is isolated if it obtains combustion air only from outdoors. An isolated CAZ doesn't require worst-case depressurization and spillage testing. However the zone must be visually inspected for connections with the home's main zone and tested for isolation.

1. Look for connections between the isolated CAZ and the home. Examples include joist spaces, transfer grilles, leaky doors, and holes for ducts or pipes.

2. Measure a base pressure from the CAZ to outdoors.

3. Perform 50-pascal blower door depressurization test. The CAZ-to-outdoors pressure should not change more than 5 pascals during the blower door test.

4. If the CAZ-to-outdoors pressure changed more than 5 pascals, perform air sealing to completely isolate the zone and retest as described above. Or alternatively perform a worst case depressurization and spillage test as described in *"Worst-Case Depressurization, Spillage, and CO" on page 212*.

7.2 SMOKE AND CARBON MONOXIDE ALARMS

All homes should have at least one smoke alarm on each level, including one near the combustion zone and at least one near the bedrooms. Carbon monoxide (CO) alarms are appropriate whenever the CO hazard is considered a likely occurrence.

Combination CO and smoke alarms are now available. Single-function alarms or combination alarms can be interconnected for whole-house protection.

Customers should be educated about the purpose and features of the alarms and what to do if an alarm sounds. Following are specifications for installing CO alarms and smoke alarms.

7.2.1 CO Alarms

CO alarms should be installed in all homes with unvented space heaters and in all homes where a furnace return air could back-draft a space heater, wood stove, fireplace, or water heater.

Observe these specifications when installing CO alarms.

- ✓ Install according to the manufacturer's instructions
- ✓ Connect to a circuit that is energized at all times
- ✓ Don't install CO alarms in these cases.
- ✓ In a room that may get too hot or cold for alarm to function properly
- ✓ Within 5 feet of a combustion appliance, vent, or chimney
- ✓ Within 5 feet of a storage area for vapor-producing chemicals
- ✓ Within 12 inches of exterior doors and windows
- ✓ Within a furnace closet or room
- ✓ With an electrical connection to a switched circuit
- ✓ With a connection to a ground-fault interrupter circuit (GFCI)

7.2.2 Smoke Alarms

Observe these specifications when installing smoke alarms.

- ✓ Install according to manufacturer's instructions.
- ✓ If mounted on a wall, mount from 4 to 12 inches from the ceiling.

✓ If mounted on a ceiling, mount at least 6 inches from the nearest wall.

✓ Connect to a circuit that is energized at all times.

Don't install smoke alarms in these situations.

✓ Within 12 inches of exterior doors and windows

✓ With an electrical connection to a switched circuit

✓ With a connection to a ground-fault interrupter circuit (GFCI)

7.3 MOISTURE PROBLEMS

Moisture causes billions of dollars worth of property damage and high energy bills each year in American homes. Water damages building materials by dissolving glues and mortar, corroding metal, and nurturing pests like mildew, mold, and dust mites. These pests, in turn, cause many cases of respiratory distress.

Water reduces the thermal resistance of insulation and other building materials. High humidity also increases air conditioning costs because the air conditioner must remove the moisture from the air to improve comfort.

The most common sources of moisture are leaky roofs and damp foundations. Other critical moisture sources include dryers venting indoors, showers, cooking appliances, and unvented gas appliances like ranges or decorative fireplaces.

Climate is also a major contributor to moisture problems. The more rain, extreme temperatures, and humid weather a region has, the more it's homes are vulnerable to moisture problems.

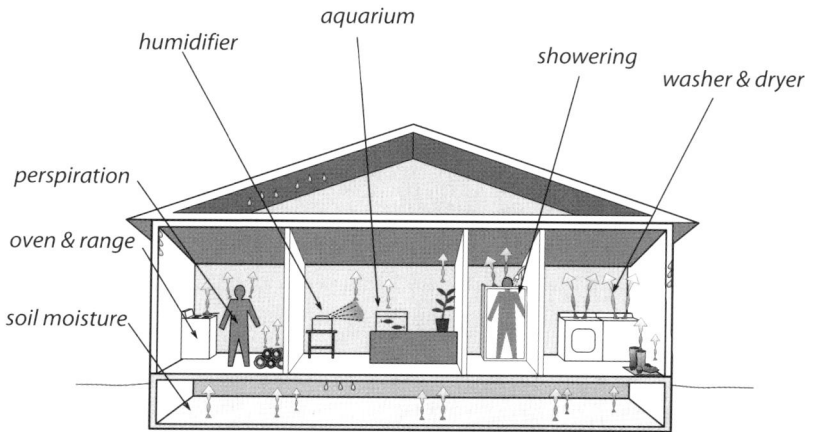

Moisture sources: Household moisture can often be controlled at the source by informed and motivated occupants.

Reducing sources of moisture is the first priority for solving moisture problems. Next most important are air and vapor barriers to prevent water vapor from migrating through building cavities. Relatively airtight homes may need mechanical ventilation to remove accumulating water vapor.

Table 7-4: Moisture Sources and Their Potential Contributions

Moisture Source	Potential Amount Pints
Ground moisture	0–105 per day
Unvented combustion space heater	0.5 –20 per hour
Seasonal evaporation from materials	6–19 per day
Dryers venting indoors	4–6 per load
Dishwashing	1–2 per day
Cooking (meals for four)	2–4 per day
Showering	0.5 per shower

7.3.1 Symptoms of Moisture Problems

Condensation on windows, walls, and other surfaces signals high relative humidity and the need to find and reduce moisture sources. During very cold weather or rapid weather changes, condensation may occur. This occasional condensation isn't a major problem. However, if window condensation is a persistent problem, reduce moisture sources, add another glass pane, or consider other remedies that lead to warmer interior surfaces. The colder the outdoor temperature, the more likely condensation is to occur. Adding insulation helps eliminate cold wall or ceiling areas where water vapor condenses.

Moisture problems arise when the moisture content of building materials reaches a threshold that allows pests like termites, dust mites, rot, and fungus to thrive. Asthma, bronchitis, and other respiratory ailments can be exacerbated by moisture problems because mold, mildew, and dust mites are potent allergens.

Rot and wood decay indicate advanced moisture damage. Unlike surface mold and mildew, wood decay fungi penetrate, soften, and weaken wood.

Peeling, blistering, or cracking paint may indicate that moisture is moving through a wall, damaging the paint and possibly the building materials underneath.

Dust mites: Biological pests create bioaerosols that can cause allergies and asthma.

Corrosion, oxidation, and rust on metal are unmistakable signs that moisture is at work. Deformed wooden surfaces may appear as damp wood swells and then warps and cracks as it dries.

Concrete and masonry efflorescence often indicates excess moisture at the home's foundation. Efflorescence is a white,

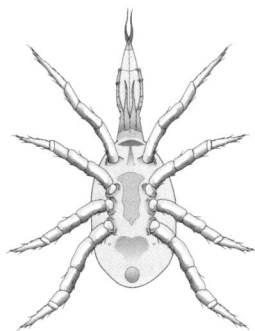

powdery deposit left by water that moves through masonry and leaves minerals from mortar or the soil behind as it evaporates.

7.3.2 Preventing Moisture Problems

Preventing moisture problems is the best way to guarantee a building's durability and its occupant's respiratory health. Follow these preventive measures before trying any of the solutions in the next section.

✓ In rainy climates, install rain gutters with downspouts that drain roof water away from the foundation. Rainwater flowing from roofs often plays a major role in dampening foundations.

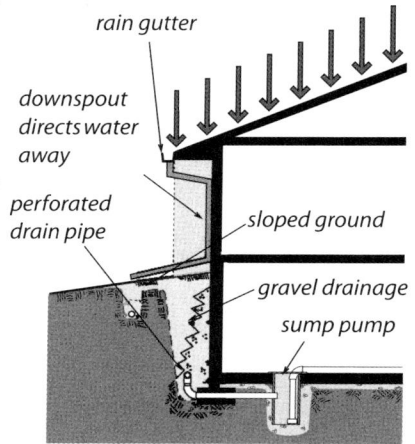

Stopping water intrusion: Take all necessary steps protect homes from water intrusion.

✓ Install a ground moisture barrier, which is a piece of heavy plastic sheeting laid on the ground. Black or clear heavy plastic film works well, but tough cross-linked polyethylene is more durable. The edges should be sealed to the foundation walls with urethane adhesive and/or mechanical fasteners. The seams should be sealed as well.

✓ A sump pump is the most effective remedy when ground water continually seeps into a basement or crawl space and collects there as standing water. Serious ground-water problems may require excavating and installing drain pipe and gravel to disperse accumulations of groundwater between a home and the underlying soil.

✓ Educate customers to avoid excessive watering around the home's perimeter. Watering lawns and plants close to the house can dampen its foundation. In wet climates, keep shrubbery away from the foundation, to allow wind circulation near the foundation.

7.3.3 Solving Moisture Problems

If moisture source reduction isn't adequate to prevent moisture problems, try these solutions after all preventive measures are in place.

✓ Install or improve air barriers and vapor barriers to prevent air leakage and vapor diffusion from transporting moisture into building cavities. *See page 70.*

✓ Add insulation to the walls, floor, and ceiling of a home to keep the indoor surfaces warmer and less prone to condensation. During cold weather, well-insulated homes can tolerate higher humidity without condensation than can poorly insulated homes.

✓ Ventilate the home with drier outdoor air to dilute the more humid indoor air. However, ventilation is only effective when the outdoor air is drier than the inside air, such as in winter. In summer, outdoor air may be more or less humid than indoor air depending on climate.

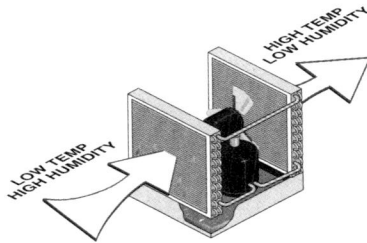

Dehumidifiers: In damp climates, dehumidifiers protect homes from excessive moisture.

✓ As a last resort, remove moisture from indoor air by cooling the air to below its dew point with compressor-based air conditioning systems in summer and dehumidifiers in winter.

7.4 OTHER POLLUTANTS

Radon and asbestos are also important hazards to both occupants and workers.

7.4.1 Radon

Radon is a dangerous indoor air pollutant that comes from the ground through rocky soil. Studies predict 16,000 lung cancer deaths per year caused by radon.

Energy conservation work probably has little effect on radon concentrations. However, all housing specialists should be aware of radon's danger, radon testing procedures, and radon mitigation strategies.

The EPA believes that any home with a radon concentration above 4 pico-Curies per liter (pC/l) of air should be modified to reduce the concentration. There are several common and reliable tests for radon, which are performed by health departments and private consultants throughout the U.S.

Since radon comes through the soil, mitigation strategies include the following.

- Installing a plastic ground barrier and carefully sealing the seams
- Sealing the walls and floor of the basement
- Ventilating the crawl space or basement to dilute radon
- Depressurizing the ground underneath the basement concrete slab

The first two mitigation strategies may already be prescribed by the weatherization work scope and may help to keep Radon levels low, after air-sealing.

7.4.2 Asbestos

Asbestos is classified as a "known carcinogen." Asbestos is found in boiler and steam-pipe insulation, floor tile, siding, roofing, and other building materials. Workers who encounter asbestos in the workplace must be trained to recognize and avoid it. Penalties for mishandling asbestos-containing materials are up to $25,000 per day.

DOE weatherization policy requires weatherization agencies to observe the following safety precautions.

- Remove asbestos siding without damaging it. Don't cut or drill asbestos siding.

- Test vermiculite for asbestos and use air monitoring if asbestos is present in the vermiculite. Don't remove vermiculite.

- Assume that asbestos is present in old grey-colored pipe and duct insulation. Don't disturb this material and caution occupants not to disturb it.

- Contract with certified asbestos testers and abatement specialists to mitigate asbestos problems before or during weatherization, if appropriate.

7.5 WHOLE-HOUSE VENTILATION SYSTEMS

This section discusses three options for design of whole-house ventilation systems.

- ✓ Exhaust ventilation
- ✓ Supply ventilation
- ✓ Balanced ventilation

Ventilation systems should always be measured and adjusted to provide only the amount of ventilation required to maintain healthy indoor air quality. Excessive ventilation wastes electricity and heating fuel.

7.5.1 Exhaust Ventilation

Exhaust ventilation systems use an exhaust fan to remove indoor air, which is replaced by infiltrating outdoor air. Better air distribution is achieved by using a remote fan that exhausts air from several rooms through small (3- to-4 inch) diameter ducts.

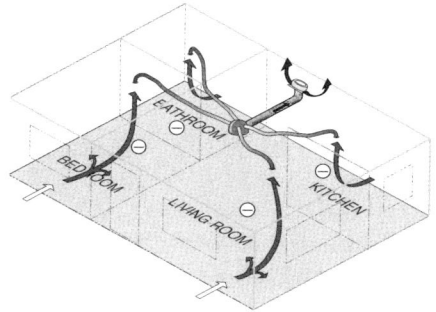

Multi-port exhaust ventilation: A multi-port ventilator creates better fresh-air distribution than a single central exhaust fan.

In smaller homes, it is adequate to install a high-quality ceiling exhaust fan, preferably in a central location. For simplicity, the central ventilation fan should run continuously, and local exhaust fans should be employed as usual, to remove moisture and odors as needed. Continuous ventilation allows for a small fan size that will minimize the depressurization compared to intermittent ventilation with a larger fan.

To replace a bathroom exhaust fan with an exhaust ventilation fan, the new fan should run continuously on low speed for whole-house ventilation. This fan should also have a high speed that occupants can use to remove moisture and odors from the bathroom quickly. Some fans have occupant sensors that switch the fan from low to high speed.

Exhaust systems create negative pressure within the home, drawing air in through leaks in the shell. This keeps moist indoor air from traveling into building cavities, reducing the likelihood of moisture accumulation in cold climates during the winter months. In hot and humid climates, however, this depressurization can draw outdoor moisture into the home. Therefore pressurized ventilation is recommended for mixed and hot climates.

Fan Specifications

Continuous ventilation is highly recommended because it simplifies design and control and also minimizes depressurization by allowing selection of the minimum-sized fan. Exhaust fans, installed as part of weatherization or home-performance work must vent to outdoors and include the following.

Specifying exhaust fans: Specify quiet energy-efficient fans.

1. The ENERGY STAR® label.

2. A weatherproof termination fitting on the roof, wall or soffit.

3. A backdraft damper, installed in the fan housing or termination fitting.

4. Noise rating and ventilation efficacy as specified.

Table 7-5: Fan Capacity, Maximum Noise Rating, & Efficacy

Fan Capacity	Noise Rating (sones)	Efficacy cfm/Watt
<50 CFM	<1 sone	Š2.8
50–100 CFM	<1.5 sones	Š2.8
>100 CFM	<2.0 sones	Š2.8

See "Duct Sizing Specs for ASHRAE Standard 62.2 - 2010" on page 262.

7.5.2 Supply Ventilation

Supply ventilation uses your furnace or heat pump as a ventilator. A 5-to-10 inch diameter duct is connected from outdoors to the furnace's main return duct. This outdoor-air supply duct often has a motorized damper that opens when the air-handler blower operates. This outdoor air is then heated or cooled by the furnace before its delivery into the living space.

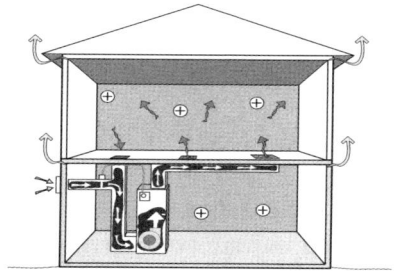

Supply ventilation: A furnace or heat pump is used for ventilation with a control that ensures sufficient ventilation.

At least one manufacturer makes a control for running the furnace blower and damper for ventilation. The control activates both the damper and blower as necessary to provide sufficient ventilation.

Supply ventilation, using the home's air handler, is never be operated continuously as with exhaust ventilation because the furnace or heat-pump blower is too large and would over-ventilate the home and waste electrical energy. Supply ventilation may not be appropriate for tight homes in very cold climates because supply ventilation can push indoor air through exterior walls, where moisture can condense on cold surfaces.

7.5.3 Balanced Ventilation

Balanced ventilation systems exhaust stale air and provide fresh air through a ducted distribution system. Of all the ventilation schemes, they do the best job of controlling pollutants in the home.

Balanced systems move equal amounts of air into and out of the home. Most balanced systems incorporate heat recovery ventilators that reclaim some of the heat and/or moisture from the exhaust air stream.

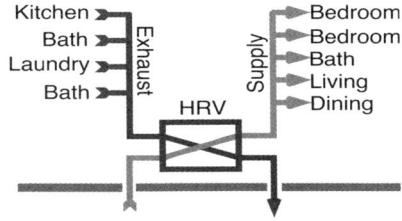

Centralized balanced ventilation: Air is exhausted from areas most likely to contain pollution and fresh air is supplied to living areas.

These complex systems can improve the safety and comfort of home, but a high standard of care is needed to assure that they operate properly. Testing and commissioning is vital during both the initial installation and periodic service calls.

Heat- and Energy-Recovery Ventilators

The difference between heat recovery ventilators (HRVs) and energy recovery ventilators (ERVs) is that HRVs transfer heat only, while ERVs transfer both sensible and latent heat (moisture) between airstreams.

Heat recovery ventilators are often installed as balanced whole-house ventilation systems. The HRV core is an air-to-air heat exchanger in which the supply and exhaust airstreams pass one another exchanging heat without mixing.

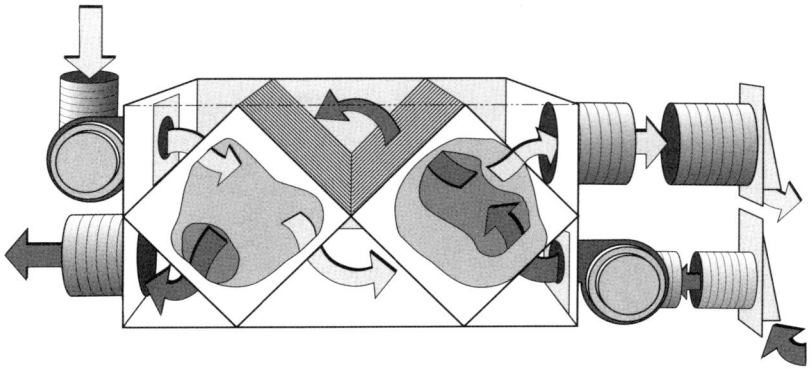

Heat-recovery ventilator: Heat from the exhaust air heats a plastic or aluminum heat exchanger, which in turn heats the fresh intake air. Two matched fans provide balanced ventilation.

7.5.4 Adaptive Ventilation

The home's residents can maintain good indoor air quality by using spot ventilation together with opening windows and doors. Depending on climate and season, residents can learn to strike a balance between clean air, comfort, and energy efficiency.

✓ Choose windows and screen doors in strategic locations to take advantage of prevailing winds.

✓ Make sure that windows and screen doors, chosen for ventilation, open and close and have effective insect screens.

✓ Open windows to provide make-up air when an exhaust fan or the clothes dryer is operating.

✓ Understand that dust and pollen may enter through windows or screen doors and consider the consequences.

7.6 LEAD-SAFE PROCEDURES

In 2010, The Environmental Protection Agency's (EPA) Lead-Safe Renovation, Repair, and Painting (RRP) rule became a legal mandate for weatherization work. Prior to the adoption of this rule, the Weatherization Assistance Program wrote rules known as: Lead-Safe Weatherization (LSW). Both the RRP and LSW approaches are covered here.

Lead dust is dangerous because it damages the neurological systems of people who ingest it. Children are often poisoned in pre-1978 homes because of paint disturbance during home improvement and because hand-to-mouth behavior is common.

Lead paint was commonly used in homes built before 1978. Technicians working on these older homes should either assume the presence of lead paint or perform tests to rule out its presence.

7.6.1 EPA RRP Requirements

The RRP rule requires lead-safe containment procedures whenever workers disturb painted surfaces of more than 6 square feet of interior surface per room or more than 20 square feet of exterior surface per side by cutting, scraping, drilling, or other dust-creating activities in pre-1978 homes. Disturbing paint on windows always requires containment.

The RRP requires certifications, warnings, dust-prevention, dust collection, and housecleaning as summarized here.

- ✓ With pre-1978 homes, either test for lead-based paint or assume that lead-based paint is present.

- ✓ Every pre-1978 weatherization or renovation job must be supervised by a certified renovator with 8 hours or EPA-approved training.

- ✓ Renovation firms must be registered with the EPA and employ one or more certified renovators.

✓ Signs and barriers must warn occupants and passersby not to enter the work area.

✓ Floor-to-ceiling dust-tight barriers must prevent the spread of dust from the work area.

Protective sheeting: Dust-tight floor-to-ceiling barriers must separate work areas from living areas, according to EPA's RRP rule.

✓ Plastic sheeting must protect surfaces and fixtures within the work area.

✓ Workers must clean work surfaces sufficiently to pass an EPA-approved dust-wipe test, conducted by the certified renovator.

✓ Workers must not track dust from the work area into the home.

7.6.2 Lead-Safe Weatherization (LSW)

Lead-Safe Weatherization (LSW) is a set of procedures developed prior to the enactment of the RRP rule. LSW requires the same basic procedures as RRP in pre-1979 homes, when disturbing more than 6 square feet of paint indoors, 20 square feet of paint outdoors, or any paint on a window. When engaging in the paint-disturbing weatherization activities, take the following precautions.

- ✓ Wear a tight-fitting respirator to protect yourself from breathing dust or other pollutants.

- ✓ Confine your work area within the home to the smallest possible floor area. Seal this area off carefully with floor-to-ceiling barriers made of disposable plastic sheeting, sealed at floor and ceiling with tape.

- ✓ Don't use heat guns or power sanders in LSW work.

- ✓ Spray water on the painted surfaces to keep dust out of the air during drilling, cutting, or scraping painted surfaces.

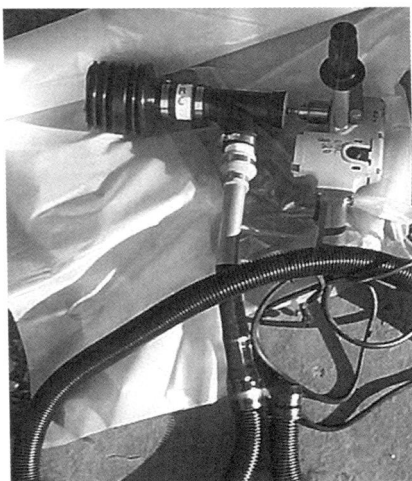

- ✓ Erect an effective dust-containment system outdoors to prevent dust contamination to the soil around the home.

- ✓ Use a dust-containment system with a HEPA vacuum when drilling holes indoors.

Drill shroud connected to HEPA vacuum: Collect dust where you're generating it.

- ✓ Avoid taking lead dust home on clothing, shoes, or tools. Wear boot covers while in the work area, and remove them to avoid tracking dirt from the work area to other parts of the house. Wear disposable coveralls, or vacuum cloth coveralls with a HEPA vacuum before leaving the work area.

- ✓ Wash thoroughly before eating, drinking, or quitting for the day.

7.6.3 Electrical Safety

Electrical safety is a basic housing need affecting home weatherization and repair. Observe the following specifications for electrical safety in weatherizing existing homes.

Non contact voltage tester: Test voltage wires near your work area and take action to turn of the circuit if appropriate.

- ✓ Whenever working around wiring, use a non-contact voltage tester to determine whether circuits are live. Turn circuits off at circuit breakers as appropriate.

S-type fuse: An S-type fuse prohibits residents from oversizing the fuse and overloading an electrical circuit.

- ✓ Inspect wiring, fuses, and circuit breakers to ensure that wiring isn't overloaded. Install S-type fuses where appropriate to prevent circuit overloading. Maximum ampacity for 14-gauge wire is 15 amps and for 12-gauge wire is 20 amps.

- ✓ Encourage your customers to replace knob-and-tube wiring.

- ✓ Never cover knob-and-tube wiring with insulation. Replace the wiring if at all possible. Construct shields to maintain a 3-inch clearance between insulation and knob-and-tube insulation if the obsolete wiring will remain.

- ✓ Confirm that all wire splices are enclosed in electrical junction boxes. If you plan to cover a junction box with insulation, mark its location with a sign or flag.

- ✓ Don't allow metal insulation shields to contact wiring.

✓ Confirm that the electrical system is grounded to either a ground rod or to a water pipe that has an uninterrupted electrical connection to the ground.

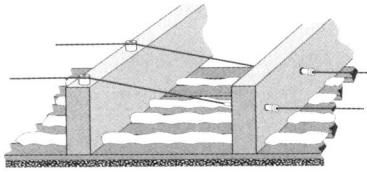

Knob and tube wiring: Obsolete and worn wiring should be replaced during energy retrofit work so that building cavities can be sufficiently insulated.

✓ Install S-type fuses where appropriate to prevent occupants from installing oversized fuses.

✓ Perform a voltage-drop test to evaluate the size and condition of hidden wiring on older homes.

7.7 WORKER HEALTH AND SAFETY

The personal health and safety of each employee is vitally important to every company. Injuries are the fourth leading cause of death in the United States, while long-term exposure to toxic materials contributes to sickness, absenteeism, and death of workers. Both of these risk factors are present during weatherization work.

Workplace safety standards have been established by the Occupational Safety and Health Administration (OSHA) and by construction trade organizations; these should be observed by weatherization staff and their contractors. Safety always has priority over other factors affecting weatherization operations.

Some hazards merit the attention because of their statistical importance. Be aware of these most common workplace hazards.

✓ Vehicle accidents.

✓ Falls.

✓ Back injuries.

- ✓ Exposure to hazardous materials.
- ✓ Electrical hazards.
- ✓ Repetitive stress injuries.

7.7.1 Commitment to Safety

It is easy to become complacent about jobsite health and safety if it is not continually emphasized. Do everything possible to create a safe work place by following these practices.

Safety education: Safety meetings are an essential part of a successful safety program.

- ✓ Arrange regular health and safety training.
- ✓ Conduct regular safety meetings.
- ✓ Keep equipment in good condition.
- ✓ Observe all state and federal standards relating to worker health and safety.

Safety requires communication and action. To protect yourself from injury and illness, learn to recognize hazards, communicate with co-workers and supervisors, and take action to reduce or eliminate hazards.

7.7.2 New Employees

New hire: New hires are several times more likely to be injured than are experienced workers.

New employees are several times more likely to injure themselves on the job compared to experienced workers. Before their first day on the job, new employees should learn about safety basics such as proper lifting, safe ladder usage, and safe operation of the power tools they will use on the job.

Be sure to inform new employees about hazardous materials they may encounter on the job. Show new hires the Material Safety Data Sheets (MSDS) required by OSHA for each material.

New employees should be required to use this common safety equipment.

- ✓ Proper clothing.
- ✓ Gloves.
- ✓ Respirators.
- ✓ Safety glasses.
- ✓ Hearing protectors.

Alcohol and drugs should be banned from the job. Staff members should be encouraged to refrain from smoking and to stay physically fit.

7.7.3 Driving

According to the Bureau of Labor Statistics, one-third of all occupational fatalities in the United States occur in motor-vehicle accidents. Staff members should organize their errands and commuting to the job site so as to minimize vehicle travel.

Safe vehicles: Maintain vehicles in good repair. Drivers and passengers should always wear seat belts.

Vehicles should be regularly inspected and repaired if necessary. These safety components are most important.

- ✓ Brake system
- ✓ Steering system
- ✓ Horns
- ✓ Headlights
- ✓ Directional signals
- ✓ Backup lights and signals

Always wear seat belts. When traveling to the job, tools and materials should be properly stowed and secured in the cargo area to prevent shifting.

7.7.4 Lifting and Back Injuries

Back injuries account for one out of every five workplace injuries. Most of these injuries are to the lower back and are the result of improper lifting, crawling in tight spaces, and using heavy tools.

Workers often injure their backs by lifting heavy or awkward loads improperly or without help. Use proper lifting techniques such as lifting with the legs and keeping a straight back whenever possible. To avoid back injury, get help before trying to lift heavy or awkward loads, stay in good physical condition, and control your weight through diet and exercise.

Workers with limited lifting abilities because of weakness or prior injury should avoid heavy lifting.

These policies help prevent jobsite injuries.

- ✓ Redesign work activities: adapt equipment and to minimize awkward movements on the job site.

- ✓ Perform strength-testing of workers, set lifting limits, and provide training for all workers on the causes and prevention of back injuries.

- ✓ Enforce breaks to prevent workers from being in straining positions for long time periods.

Awkward loads: Ask for help when moving heavy or awkward loads.

- ✓ Share the difficult work among capable crew members.

7.7.5 Respiratory Health

Wear a respirator when working in a dusty environment. Common construction dust can contain of toxins including lead, asbestos, and chemicals. Drilling, cutting, scraping can stir up toxic dust which may then be inhaled.

Test your respirators to be sure they have a good fit.

- ✓ Check the straps and face piece to be sure they are soft and free of cracks.

- ✓ Strap on the respirator and adjust the straps to be snug but comfortable.

- ✓ Close the exhalation valve with a hand.

- ✓ Exhale gently and check for leaks around the edges.

- ✓ If there are leaks, adjust or repair the respirator.

Workers with beards, facial scars, and thick temple bars on eyeglasses must take special care to get a good seal.

Some airborne hazards require specialized respirator filters. Workers spraying polyurethane foam, for example, should use a respirator canister designed to filter organic vapors, and they should ventilate the area where the foam is being sprayed. For areas like crawl spaces that are difficult to ventilate, workers should use a supplied-air, positive-pressure respirator.

Learn how to recognize asbestos insulation that may be installed around older furnaces and boilers.

Control dust in your client's homes by erecting temporary barriers when you are doing work that may release toxic dust. Wear coveralls when entering attics or crawl spaces. Coveralls should be disposable or laundered professionally.

7.7.6 Hazardous Materials

Your health and safety can be threatened by hazardous materials used on the job. Workers often fail to protect themselves from hazardous materials because they don't recognize and understand their health effects. Breathing hazardous materials, absorbing them through the skin, and coming into eye contact with hazardous materials are common ways workers are affected.

OSHA regulations require employers to notify and train employees about hazardous materials used on the job. A Material Safety Data Sheet (MSDS) for every workplace hazardous material should be readily available to employees. Copies of MSDSs are obtained from manufacturers or their distributors. Employees should know where MSDSs are kept and how to interpret them.

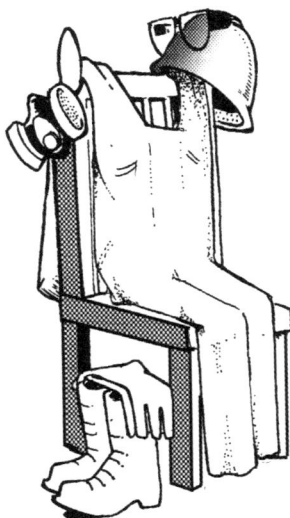

Personal protective equipment: Employees should own and maintain protective equipment to protect themselves from hazardous materials.

Learn how to avoid exposure to hazardous materials used on the job and how to clean up chemical spills. Employees should be instructed to use the appropriate protective equipment that is recommended by the MSDS.

7.7.7 Falls

Falls off ladders and stairs cause 13% of workplace injuries according to the National Safety Council. Falls from the same elevation such as slips and trips account for approximately 7% of workplace injuries.

Broken ladders, and ladders that slip because they haven't been anchored properly, are both major causes of on-the-job falls. Step ladders, for instance, are often used for work that is too far off the ground, forcing workers to stand on the top step or to reach too far.

Ladders: Ladders are the most dangerous tools workers use.

OSHA regulations include these important guidelines for ladder use.

- ✓ All ladders should be kept in good repair, and should be replaced if they have missing steps or cracked side-rails.

- ✓ Broken ladders should be removed from the equipment storage area.

- ✓ Extension ladders should be set to extend at least three feet above the area they access.

- ✓ Ladders shouldn't have a pitch steeper than four feet of rise for each foot the base is away from the building.

- ✓ Ladders must be blocked or tied firmly in place at the top and bottom if the above rule cannot be observed.

- ✓ Metal ladders should not be used where they may come in contact with electrical conductors.

- ✓ Ladders must be maintained free of oil, grease, and other slipping hazards.
- ✓ Ladders must not be loaded beyond the maximum load for which they were built.
- ✓ Workers should avoid carrying heavy loads up ladders and operating power tools from ladders.

Scaffolding must be used when working above-ground for sustained time periods. Scaffolds should be built plumb and level. Each leg should be stabilized so that it supports equal weight as other legs. This is especially important on unlevel ground. Planks should be secured to the structure and handrails provided on the sides and ends of the walkway.

Good housekeeping: Clear stairs and walkways are essential to protect workers and clients alike from falls.

Workplaces should be policed regularly to remove slipping and tripping hazards. Workers carrying loads should establish a debris-free walkway.

7.7.8 Tool Safety

The tools used in construction work are dangerous if used improperly. About 90,000 people hurt themselves with hand tools each year. One moment of inattention can cause an injury that can change a worker's life.

These basic safety rules can reduce the hazards of using hand and power tools.

Electrical safety: Cords should be maintained in good condition. Ground-fault-interrupter cords or outlets should be used in wet conditions.

✓ Use the right tool for the job.

✓ Keep all tools in good condition with regular maintenance.

✓ Inspect tools for damage before using them.

✓ Operate tools according to the manufacturer's instructions.

✓ Use appropriate personal protective equipment.

✓ Use ground-fault-interrupter outlets or extension cords.

✓ Repetitive Stress Injuries

✓ Repetitive stress injuries are caused by over-working certain parts of your body. Poor body posture, such as reaching above your head when operating a power drill, can encourage these injuries. Good work habits help prevent this type of injury.

✓ Use a comfortable arm and hand posture when operating tools for a long period of time.

✓ Change the angle and location of your work surface frequently.

- ✓ Mix your difficult tasks with easier ones.

- ✓ Carry smaller loads.

- ✓ Take short rest breaks periodically, and stretch any tight muscles during this time.

When you purchase hand and power tools, look for models with ergonomic designs that place less stress on your body.

Health and Safety

Appendices

A-1 Resources

ASHRAE 2005 Fundamentals, American Society of Heating, Refrigerating and Air Conditioning Engineers, Inc; Atlanta GA 1993, www.ashrae.org

ASHRAE Standard 62.2 2010, American Society of Heating, Refrigerating and Air Conditioning Engineers, Inc; Atlanta GA 2010, www.ashrae.org

Homeowner's Handbook to Energy Efficiency, Saturn Resource Management, by John Krigger and Chris Dorsi, www.srmi.biz

HVAC Duct Standard, Building Performance Institute (BPI), draft, www.bpi.org

HVAC Quality Installation Specification, ACCA/ANSI 5 QI-2007, Air Conditioning Contractors of America, www.acca.org

International Energy Conservation Code 2009, International Code Council, www.iccsafe.org

International Residential Code 2009, International Code Council, www.iccsafe.org

Residential Energy: Cost Savings and Comfort for Existing Buildings, Saturn Resource Management, by John Krigger and Chris Dorsi, Fourth Edition, 2004, www.srmi.biz

Residential Weatherization Specification Manual, Northwest Regional Technical Forum, 2011, www.nwcouncil.org/energy/rtf/

Standardized Work Specifications for Residential Energy Efficiency Retrofits, DOE EERE Weatherization and Intergovernmental Program, http://www1.eere.energy.gov/wip/wap.html

A–2 REQUIRED DIAGNOSTIC EQUIPMENT

Minimum Equipment For Instrumented Air Sealing

- ✓ Fully instrumented and calibrated blower door.
- ✓ Two-channel handheld digital manometer.
- ✓ Smoke generating equipment.

Minimum Equipment for Heating System Analysis

- ✓ Combustion analyzer.
- ✓ CO testing capacity.
- ✓ Draft gauge or manometer.
- ✓ Heat exchanger leakage testing equipment.
- ✓ Ammeter (sensitive enough to adjust thermostat anticipators).
- ✓ Combustible-gas leak detector.

A-3 Tools for Air Sealing and Insulating

Insulation blower	Broom and dust pan
Blower hoses 4, 3, 2.5, & 2 inch	Cat's paw
Fill tubes and hose fittings	Caulking guns
Coveralls and gloves	Chisels: cold and wood
First-aid kit	Cleaning fluid and rags
Hard hat	End nippers
Respirators and filters	Flashlight
Safety glasses	Hack saw and blades
Ext. ladders, leveler, & hooks	Hammers and wrecking bars
Portable lights	Hand saws
Scaffold, planks, and handrail	Hand staplers
Step ladders: 4, 6, & 8 foot	Metal & Vinyl-siding zip tools
Circular saw with blades	Mirror
Compressor and power stapler	Pliers: electrical & slip-joint
Drill index with bits	Putty knives and scrapers
Drills, drivers, and bits	Putty warmer
Extension cords	Scratch awl and pin punches
HEPA vacuum with attachments	Steel tape measures
Lead paint drill shroud	Screw drivers and nut drivers
Reciprocating saw with blades	Squares: frame, combo, drywall
Shop vacuum, hoses, attachments	Tin snips: hand and electric

A–4 Materials for Air Sealing and Insulating

Cellulose insulation	Drywall
Closed-cell foam tape	Compact fluorescent lamps
Fiberglass batts	Energy-saving shower heads
Fiberglass blowing wool	Programmable thermostats
Fiberglass duct wrap	Window weatherstrips
Foam backer rod	High-quality door weatherstrip
Foam pipe sleeves	Replacement refrigerators
One-part squirt foam	Replacement water heaters
Sheet foam insulation	1/4-inch plywood or hardboard
Two-part spray foam	Assorted lumber
Water heater insulation	Assorted screws and nails
Assorted chimney pipe	Assorted staples
Assorted furnace filters	Construction adhesive
Duct mastic and web tape	Disposable coveralls, boot cov-
Duct tape and electrical tape	Disposable paint brushes
Furnace filter material	Plastic garbage bags
Proper vents	Plastic sheeting
Galvanized sheet metal	Pop riveter
Acoustical sealant	Putty tape
Bronze v-seal weatherstrip	Silicone or urethane caulk
Jamb-up weatherstrip	Siliconized acrylic-latex caulk
Customer-education booklets	Portable tape recorder

A–5 R-VALUES FOR COMMON MATERIALS

Material	R-value
Fiberglass or rock wool batts and blown 1″	2.8–4.0
Blown cellulose 1″	3.0–4.0
Vermiculite loose fill 1″	2.7
Perlite 1″	2.4
White expanded polystyrene foam (beadboard) 1″	3.9–4.3
Polyurethane/polyisocyanurate foam 1″	6.2–7.0
Extruded polystyrene 1″	5.0
Sprayed 2-part polyurethane foam 1″	5.8–6.6
Icynene foam 1″	3.6
Oriented strand board (OSB) or plywood $^1/_2$″	1.6
Concrete or stucco 1″	0.1
Wood 1″	1.0
Carpet/pad $^1/_2$″	2.0
Wood siding $^3/_8$–$^3/_4$″	0.6–1.0
Concrete block 8″	1.1
Asphalt shingles	0.44
Fired clay bricks 1″	0.1–0.4
Gypsum or plasterboard $^1/_2$″	0.4
Single pane glass $^1/_8$″	0.9
Low-e insulated dbl. pane unit glass (Varies according to Solar Heat Gain Coefficient (SHGC) rating.)	3.3–4.2
Triple glazed glass with 2 low-e coatings	8.3

A–6 MINIMUM R-VALUES FROM THE IECC

R-Values of Building Assemblies from the IECC

Climate Zone	Ceiling	Wood Wall[1]	Masonry Wall[2]	Floor	Basement Wall[3]	Crawl Space Wall[3]
1	30	13	3/4	13	0	0
2	30	13	4/6	13	0	0
3	30	13	5/8	19	5/13	5/13
4 except marine	38	13	5/10	19	10/13	10/13
5 and marine 4	38	20 or 13 + 5	13/17	30	10/13	10/13
6	49	20 or 13 + 5	15/19	30	15/19	10/13
7 & 8	49	21	19/21	38	15/19	10/13

The values here are from the 2009 International Energy Conservation Code (IECC) and are minimum required values.

1. "13 + 5" means R-13 plus R-5 insulated sheathing.
2. Or insulation sufficient to fill framing cavity (R-19 minimum).
3. "13/17" means R-15 continuous interior or exterior insulated sheathing or R-19 interior cavity wall insulation.

A–7 CALCULATING ATTIC INSULATION

Auditors and inspectors also help crews determine how much insulation is needed for ceilings and walls.

Calculating Attic Loose-Fill Insulation

Loose-fill attic insulation should be installed to a uniform depth to attain proper coverage (bags per square foot) so it attains the desired R-value at the settled thickness. Follow the manufacturer's labeling in order to achieve the correct density to meet the required R-value. Attic insulation always settles: cellulose settles between 10% to 20% and fiberglass settles between 3% to 10%. For this reason, it's best to calculate insulation density in square feet per bag rather than installed thickness.

Insulation Calculation Table

R-Value at 75° F mean Temperature	Minimum Thickness	Maximum Net Coverage	
Desired R-Value of Insulation	Minimum Insulation Depth	Maximum Coverage per Bag (sq. ft.)	Bags per 1000 sq. ft.
R-60	16.0	11.7	85.8
R-50	13.3	14.0	71.5
R-44	11.7	15.9	62.9
R-40	10.7	17.5	57.2
R-38	10.1	18.4	54.4
R-32	8.5	21.6	45.8
R-30	8.0	23.3	42.9
R-24	6.4	29.1	34.3
R-22	5.9	31.8	31.5
R-19	5.1	36.8	27.2
R-13	3.5	53.8	18.6
R-11	2.9	63.6	15.7

Insulation Coverage Table: This table is provided by Weather Blanket Corporation. Coverage and other insulation characteristics will vary from manufacturer to manufacturer.

Example: Calculating Number of Bags

30 FT X 50 FT= 1500 SQ FT

Width   Length   Area of Attic

Step 1: Calculate area of attic
Multiple length times width of the attic to get the area of attic.

R-50 – R-26 = R-24

Desired R   Existing R   R Needed to Add

Step 2: Calculate R-value that you need to add
Subtract existing R from desired R to get the R-value you need to add.

1500 SQ FT ÷ 29.1 = 52 BAGS

Net wall   Sq. Ft.   Estimated Bag
Area   Coverage per   Count
Bag
(from chart)

STEP 3: Calculate bag count
Divide area of attic by coverage per bag from the chart on the bag (number highlighted in chart on *"Insulation Calculation Table" on page 254*) to get your Estimated Bag Count.

Example: Calculating Density of Attic Insulation

1500 SQ FT X 6.4/12 FT= 800 CU FT

Area   Depth in Inches   Inches per Foot   Volume of Insulation

Step 1: Calculate volume of installed insulation
Multiple area times depth of the attic insulation to get the volume of insulation.

52 BAGS X 24 LBS/BAG = 1248 LBS

Number of Bags   Weight of a Bag   Installed Weight

Step 2: Calculate the weight of insulation you installed
Take the number of bags times the weight per bag to get the total weight.

1248 LBS ÷ 800 CU FT= 1.56 LBS/CU FT

Pounds of Insulation   Insulation Volume   Installed Density

STEP 3: Calculate density of installed insulation
Divide pounds of insulation by cubic feet of insulation volume to get density.

Note
Density should be between 1.3 and 2.0 pounds per cubic foot or conform to manufacturer's specifications for density, coverage, and bag count for the desired R-value.

A–8 CALCULATING WALL INSULATION

Wall insulation should be installed to a density of 3.5 to 4.5 pounds per cubic foot. These calculations serve to calculate the number of bags necessary to insulate walls and to judge density after completing the wall-insulation job. Calculate the bag count based on information from the agency's insulation supplier.

Example: Calculating Number of Bags for Wall Insulation

$$(2 \times 50 \text{ FT}) + (2 \times 30 \text{ FT}) = 160 \text{ FT}$$

Length   Width   Perimeter of House

STEP 1: Calculate perimeter of house
Calculate the perimeter of the house. If the house is a simple rectangle or near a simple rectangle, use the formula above. If the house has numerous unequal sides, simply add the lengths together to find the perimeter.

$$160 \text{ FT} \times 8 \text{ FT} = 1280 \text{ SQ FT}$$

Perimeter of House   Height of Wall   Total Wall Area

STEP 2: Calculate total wall area
After calculating the perimeter of the house, multiply it times the wall height. This will give you the total wall area.

$$1280 \text{ SQ FT} - 150 \text{ SQ FT} = 1130 \text{ SQ FT}$$

Total Wall Area   Area of Windows and Doors   Net Wall Area

STEP 3: Calculate net wall area
Calculate the sum of the areas of windows and doors. Subtract them from the total wall area to get net wall area.

$$\frac{1130 \text{ SQ FT} \times 1.2 \text{ LBS/SQ FT}}{24 \text{ LB PER BAG}} = 57 \text{ BAGS}$$

Net Wall Area   Pounds per Square Foot   Bags of Insulation Needed

Weight of a Bag

STEP 4: Calculate bag count
To achieve 4.0 lbs. per cubic foot, multiply net wall area by 1.2 pounds per square foot for a 2-by-4 wall (4.0 lbs. per cubic foot ÷ 12 x 3.5 = 1.2). Then divide by the number of pounds per bag to get the bag count.

Example: Calculating Density of Wall Insulation

$$1280 \text{ SQ FT} \times 3.5/12 \text{ FT} = 373 \text{ CU FT}$$

Net Wall Area   Inches of Wall Depth   Inches per Foot   Wall Volume

STEP 1: Calculate wall volume
Multiply the wall's surface area times the depth on the wall cavity converted to feet.

57 BAGS X 24 LBS/BAG = 1368 LBS

Bags Installed

Weight of a Bag

Pounds of Insulation

STEP 2: Calculate weight of insulation
Multiply number of bags you installed times the weight of a single bag to get the weight of the installed insulation.

1388 LBS ÷ 373 CU FT= 3.67 LBS/CU FT

Pounds of
Insulation

Insulation
Volume

Installed Density

STEP 3: Calculate density of installed insulation
Divide pounds of insulation by cubic feet of insulation volume to calculate density.

A–9 CALCULATING MOBILE HOME INSULATION

Consider a 14′ x 66′ mobile home, totaling 924 square feet.

- ✓ Ceiling: 9″ cavity at the center and 2″ cavity at the edge with a 2″ batt

- ✓ Belly: $5^1/_2$″ cavity at the wings and $16^1/_2$″ cavity at the center with a 2″ batt fastened to floor bottom

- ✓ Walls: $3^1/_2$″ cavity with a $1^1/_2$″ batt at $7^1/_2$′ high

General formulas

CAVITY VOLUME X DESIRED DENSITY = WEIGHT OF INSULATION

WEIGHT OF INSULATION ÷ POUNDS PER BAG = BAGS OF INSULATION

Ceiling Bag Count Estimates

1. Calculate the average ceiling cavity (9″ + 2″ = 11″) (11″ ÷ 2 = $5^1/_2$″ average cavity)

2. ($5^1/_2$″ cavity minus the 2″ batt = $3^1/_2$″ cavity). The existing insulation batt will compress when additional insulation is added, allow 1″ for compression ($3^1/_2$″ + 1″ = $4^1/_2$″ cavity)

3. Convert $4^1/_2$″ to feet (4.5″/12″ = 0.375′)

4. Multiply 0.375′ x 924 sq. ft. = 346.5 cubic feet

5. Multiply cubic feet by desired density: Fiberglass ceiling insulation density must be 1.0 to 1.5 lbs/cubic foot.

   a. 347 x 1.0 = 347 lbs. / 35(lbs/bag) = 9.9 bags

   b. 347 x 1.25 = 434 lbs. / 35(lbs/bag) = 12.4 bags

   c. 347 x 1.5 = 521 lbs. / 35(lbs/bag) = 14.9 bags

Belly Bag Count Estimates

Calculate the average belly cavity $(5\text{-}^1/_2'' + 16\text{-}^1/_2'' = 22'')$

$(22'' \div 2 = 11''$ average cavity) $(11''$ cavity $- 2''$ batt $= 9''$ cavity) The existing insulation batt will compress when additional insulation is added, allow $1''$ for compression $(9'' + 1'' = 10''$ cavity)

1. Convert $10''$ to feet $(10'' / 12'' = 0.83')$

2. Multiply $0.83' \times 924$ square feet $= 767$ cubic feet

3. Multiply cubic feet by desired density.

4. Belly insulation density at 1.0 to 1.5 lbs/cubic foot.

   a. $767 \times 1.0 = 767$ lbs. $/ 35$(lbs/bag) $= 22$ bags

   b. $767 \times 1.25 = 959$ lbs. $/ 35$(lbs/bag) $= 27$ bags

   c. $767 \times 1.5 = 1151$ lbs. $/ 35$(lbs/bag) $= 33$ bags

A–10 Duct Sizing Specs for ASHRAE Standard 62.2 - 2010

Duct Type	Flex Duct				Smooth Hard Duct			
Fan Rating in CFM	50	80	100	125	50	80	100	125
Duct diameter	Maximum Duct Length in feet							
3"	not allowed	not allowed	not allowed	not allowed	5'	not allowed	not allowed	not allowed
4"	70'	3'	not allowed	not allowed	105'	35'	5'	not allowed
5"	unlimited	70'	35'	20'	unlimited	135'	85'	55'
6"	unlimited	unlimited	125'	95'	unlimited	unlimited	unlimited	145'
7"	unlimited	unlimited	unlimited	unlimited	unlimited	unlimited	unlimited	unlimited

Table assumes no elbows. Deduct 15 ft of allowable duct length for each elbow.

Table adapted from ASHRAE Standard 62.2 (2010), page 6, Table 5.3.

TABLES AND ILLUSTRATIONS

Building Shell Basics

Diagnosing Shell & Duct Air Leakage

Air Sealing Homes

Installing Insulation

Windows and Doors

Mobile Homes

Health and Safety

INDEX

W-Z

Index